NUMERICAL ANALYSIS—
The Mathematics of Computing
VOLUME

Volume 2 contains sufficient material for the more advanced student and covers the requirements of many university courses.

NUMERICAL ANALYSIS—
The Mathematics of Computing
VOLUME 1

W. A. WATSON, B.Sc.
Head of Mathematics Department,
The Sweyne School, Rayleigh, Essex.

T. PHILIPSON, M.Sc.
Head of Mathematics Department,
Medway and Maidstone College of Technology,
Chatham, Kent.

P. J. OATES, B.Sc. Tech.
Head of Mathematics Department,
The Grove School,
Market Drayton, Shropshire.

NEW YORK

AMERICAN ELSEVIER PUBLISHING COMPANY, INC.

First Published 1969

American Elsevier Publishing Company, Inc.
52 Vanderbilt Avenue,
New York 17, New York

First Published in Great Britain by
Edward Arnold (Publishers) Ltd

Standard Book Number 444-19736-2

Library of Congress Catalog Card Number: 76-84836

Printed in Great Britain

Foreword

Mathematics plays an essential part in a wide variety of subjects. These subjects not only include all branches of Engineering and the Physical and Biological Sciences, but also such techniques as Business Management, Factory Operation, Traffic Control and Logic. Moreover, the further any of these subjects is studied, the more mathematical it is likely to become.

Mathematical studies involve the student in the solution of problem exercises so that the understanding of concepts, theories and their applications is deepened. Frequently the exercises involving numerical calculations are especially constructed so that the solutions are integers or simple rational fractions. If the exercises the student attempts are exclusively of this type it can result in the same mathematical methods, when used in the laboratory to calculate results from experiments, appearing to be very different and difficult. But in spite of the computational difficulties it is essential that the student should be able to complete calculations involving numbers which are not integers or simple rational fractions because it is calculations of these types which arise so frequently from actual real life situations.

Again at appropriate stages in his development the student should be helped to realize that the logical, step by step sequence by which his mathematical tuition progresses (ideal though this sequence may be) only covers a limited range: problems being included only if formal mathematical solutions exist or can be improvised by exercising a little ingenuity. The solution of many problems, even quite elementary ones, which arise in engineering and science depends on equations which cannot be solved analytically (i.e. by using algebra and calculus to produce a formula), and yet it often happens on account of the practical, technological or economic significance of these problems that solutions must be provided and provided quickly and accurately. The way out of this apparently impossible dilemma is to utilize suitable numerical methods to provide acceptable numerical solutions for these problems: the solutions being acceptable if they can be stated to be of an accuracy which is satisfactory in the circumstances of the problem.

This indicates, rather briefly, why it is necessary to consider methods of obtaining numerical solutions to practical problems. But the reader will find that the usefulness of these methods is not restricted to those cases in which no analytical solutions can be found. Indeed there are many cases in which an analytical solution exists, but that as this is of a somewhat involved nature it is preferable to utilize a numerical method on account of the ease with which a numerical solution may thus be obtained.

Numerical Analysis—the Mathematics of Computing Vols. I and II attempt to provide an elementary and sound introduction to some of the most important methods of obtaining numerical solutions of known accuracy to a wide variety of problems.

The study of these numerical methods implies some prior knowledge of the mathematics involved. For example, before reading Chapter 4 a student would need to be aware of the need to obtain solutions of equations and have met some of the difficulties in solving equations by the usual direct methods. Again, before studying Chapter 9 a student would need to be familiar with the meaning of an integral and to have had some experience in evaluating definite integrals of some known elementary functions.

In addition to this general mathematical background essential for a full appreciation of certain chapters, use has also been made of a few important theorems in the derivation of some of the methods and formulae. To help the beginner, brief notes on the Binomial theorem and Taylor's theorem have been included in § 1.2.1 and 1.2.2.

Volume II contains: (a) a detailed discussion of interpolation; (b) methods for numerical integration and differentiation and introductory work on the numerical solution of differential equations and (c) material on curve fitting by least squares and the summing of slowly convergent series.

There is reason to be confident that these volumes will be of real value to students and teachers in schools, colleges and universities.

E. Kerr

Paisley
1968

Preface

From practical experience the authors believe that there exists a need for a text book on elementary numerical analysis, suitable for use as an introduction to the subject. With mathematical knowledge advancing so rapidly and numerical analysis an accepted part of mathematical degree courses, there is a case for the introduction of some, if not all, of the topics we have considered into the early curriculum. Hand-calculating machines are becoming readily available and, in the future, knowledge of how to use such mechanical aids will be essential to students and technologists. The study of numerical analysis also introduces the student to mathematical ideas and techniques which, as yet, are not generally incorporated into school syllabuses.

The authors wish to acknowledge the considerable help given to them in this venture by ADDO Ltd. It was under the auspices of ADDO Ltd. that they were first introduced to each other and since that day have been able to meet together at frequent intervals to discuss and amend draft material for the book.

The authors find it difficult to express in words the great debt they owe to Dr. Kerr, Principal of Paisley College of Technology, who has acted as general editor throughout the years of preparation. Dr. Kerr has given considerable help and advice most freely and generously.

Finally our grateful thanks to Mrs. Jill Pilkington (Née Swindell) for all the secretarial help she has given in typing the manuscript.

We dedicate this book to all the students who have experienced our tuition, worked through the examples and in some cases contributed original examples themselves. The authors wish to thank the Associated Examining Board and The Cambridge Local Examinations Syndicate for permission to include questions set by them in past examination papers.

W.A.W.
T.P.
P.J.O.

1968

Contents

1

Introduction and the Use of Hand-calculating Machines

1.1 THE CHOICE OF MECHANICAL CALCULATING AIDS

Before discussing the various methods for obtaining numerical solutions to a variety of problems, it is wise to survey briefly some of the many aids which have been developed to reduce the tedium and hard work of long calculations. For in addition to choosing the most appropriate method for solving a particular problem it is important to know what calculating aids should be used to help with the resulting calculations.

Methods and calculating aids should be selected with the following criteria in mind;

(a) To ensure correct working.
(b) To ensure the desired overall accuracy or the maximum accuracy consistent with the data.
(c) To minimize the time taken.
(d) To minimize the effort required.

Real skill in numerical work lies in developing the art of avoiding unnecessary arithmetic. The smaller the amount of hard work which must be done in a calculation the fewer will be the number of mistakes made.

A short survey of some of these aids follows, with comments which may help to make clear their relative spheres of usefulness.

1.1.1 Mathematical tables

Values of well known functions have been calculated and are available in books of tables. We have made use of the following:

(a) Logarithm and other Tables for Schools by Frank Castle—Macmillan
(b) Shorter Six-figure Mathematical Tables by L. J. Comrie—Chambers
(c) Interpolation and Allied Tables by H.M. Nautical Almanac Office—H.M. Stationery Office.

Frequent references will be found to the above, and the student is recommended to read carefully the detailed instruction in the use of tables, which they all contain.

1.1.2 The slide rule is an instrument which can be used to perform calculations involving multiplication and division. Basically two equal logarithmic scales are used. Each number indicated on such a scale is situated at a distance from the beginning of the scale proportional to the logarithm of the number.

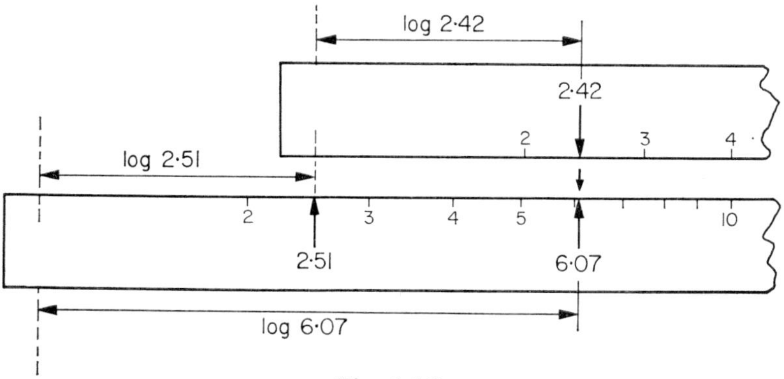

Fig. 1.1.2

The diagram illustrates the setting of the scales for the multiplication of any number by 2·51 and in particular 2·51 × 2·42 = 6·07 or 251 × 0·242 = 60·7. A combination of four log scales together with log sine and log tangent scales is commonly used, making the instrument capable of being used in a wide range of calculations.

The accuracy obtainable in this case depends on the operator's skill (and good eyesight) but is in any case limited to about 0·1 % at best, because numbers can, in general, only be set (or read off) to about three significant figures.

1.1.3 The planimeter is a mechanical integrator and is used to evaluate a given area (usually irregular in shape) by direct measurement. It is also used to evaluate a definite integral after the area representing the integral has been accurately drawn to scale. The instrument is regularly used by engineers and draughtsmen.

One common type of planimeter consists of two straight rods AB, BC freely hinged at B, Figure 1.1.3. The end A is maintained in a fixed position while the

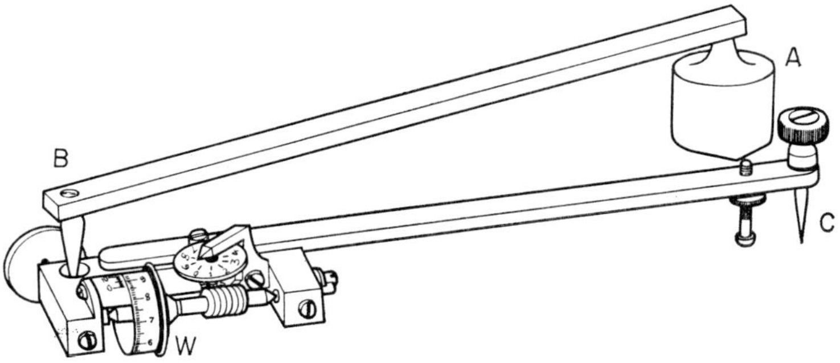

Fig. 1.1.3

pointer at the end C is moved once round the perimeter of the area to be measured and returned exactly to its starting point. The angle turned through by a wheel W, mounted on the arm BC, is recorded on a dial and is proportional to the given area.

Accuracy obtainable depends on carefully following the curve but four significant figures can be read off from the dial with the aid of the vernier scale provided.

1.1.4 Hand-calculating machines

In this type of machine a number entered in the setting register is added to any number already in the accumulator or product register by one forward turn of the handle. The number in the setting register is subtracted by one backward rotation of the handle. Multiplication is achieved by repeated addition and use of the movable carriage to give place values. Division similarly can be accomplished by repeated subtraction of the divisor. (See section §1.3 et seq. for detailed instructions.)

The accuracy obtainable on these machines depends only on the number of digits which can be accommodated in the registers. For example on an $8 \times 5 \times 13$ machine if an 8 figure number is multiplied by a 5 figure number the exact product is obtained. However, if the 5 figure multiplier is a rounded number 'correct' to five figures then, of course, only some of the 12 or 13 figures in the product are reliable. A detailed discussion of the effect of rounding errors is given below. See § 1.11, 1.14, 1.16.

1.1.5 Electrical calculating machines

The basic principle of electrical calculators is exactly the same as for hand machines, but all four fundamental operations of arithmetic are performed automatically, i.e. the required rotations and carriage movements are actuated by an electric motor. The two given numbers are set and then the appropriate press-button is used to start the required operation. There is a wide range of different models available, with a choice of capacity and additional features such as storage registers and transfer facilities. Whether the cost of a machine with these extras is justifiable will depend on the quantity and the type of calculation, which has to be done. For example in statistical work involving tabulated values of two variables x and y it is useful to have a machine large enough to form the sums $\sum x$, $\sum y$, $\sum x^2$ and $\sum xy$ simultaneously.

As with hand machines the accuracy which can be obtained depends only on the number of digits accommodated by the registers.

1.1.6 The analogue computer

An analogue computer is an instrument or machine in which the values of variables occurring in a problem are continuously represented by some convenient physical quantity.

According to this definition a slide rule is a simple analogue computer for the numbers and their product are represented by lengths. Similarly in the case of the planimeter an area is represented by the angular rotation of a wheel.

Another type of mechanical integrator is shown diagrammatically in Figure 1.1.6(a).

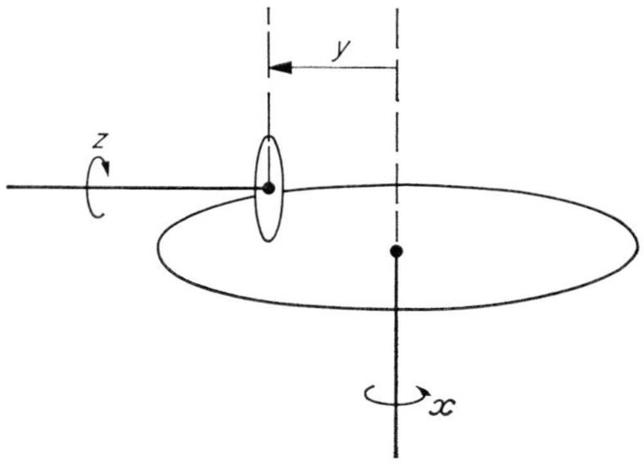

Fig. 1.1.6(a)

The angle turned through by the large horizontal wheel (driven by motor) represents the variable x. Then if the variable distance y is always kept equal to a function of x [say $f(x)$] then the smaller wheel of radius r is rotated through an angle z such that for each elementary part of the motion we have:

r. $\delta z = y. \delta x$ which is the distance moved by the point of contact of the wheels in any short interval of time, δt. (Provided of course no slipping occurs.)

$$\text{Hence } \delta z = \frac{1}{r} . y . \delta x \quad \therefore z = \frac{1}{r} \int_{x_0}^{x} f(x) . \mathrm{d}x$$

i.e. z is proportional to the integral of $f(x)$.

Electronic integrator

An electronic amplifier, maintained in action by a separate power supply, can receive an input consisting of a variable voltage. The output is a similar variable voltage increased in magnitude i.e. multiplied by constant amplifying factor (M). It can also be used to add several variable voltages, the output being M times the sum of the voltages fed in.

If a suitable capacitance (i.e. condenser) is connected in parallel with a high gain (i.e. large amplification factor) amplifier, this circuit now acts as an integrator. That is if the variable input voltage has the value $f(t)$ at any time t then the output will be proportional to $\int_{t_0}^{t} f(t)\mathrm{d}t$. This result is in theory only approximate, but is very nearly true if M is large enough*. Fig. 1.1.6(b).

*For details read:—*Fundamental Analogue Techniques*, R. J. A. Paul, Blackie and Sons, Pages 36–39.

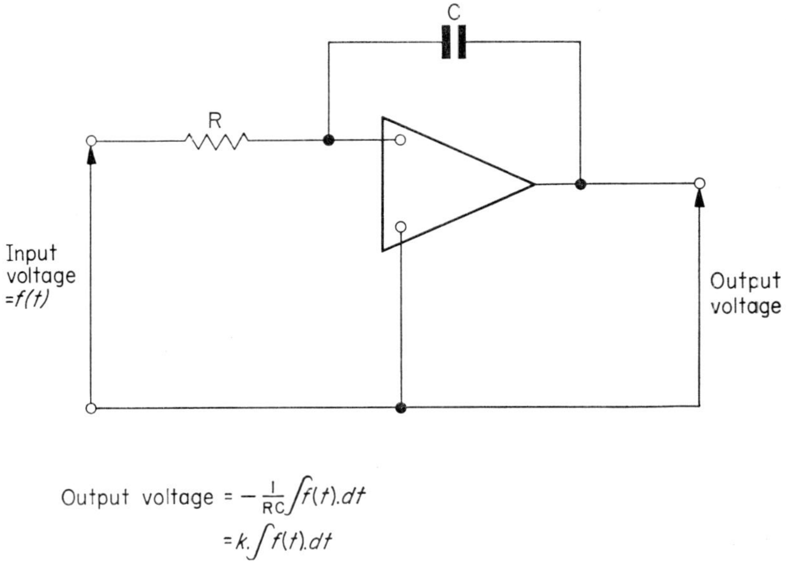

Fig. 1.1.6(b)

The Analogue Computer, as generally understood, is a combination of integrator and adder units such as described above and can be either mechanical or electronic.

Using two integrators coupled in series we could arrange them as follows: Figure 1.1.6(c) shows two electronic integrators connected in series. Multiples of the two outputs $k_1\frac{dy}{dt}$ and k_1k_2y are added and their sum a$\frac{dy}{dt} + by$ is passed

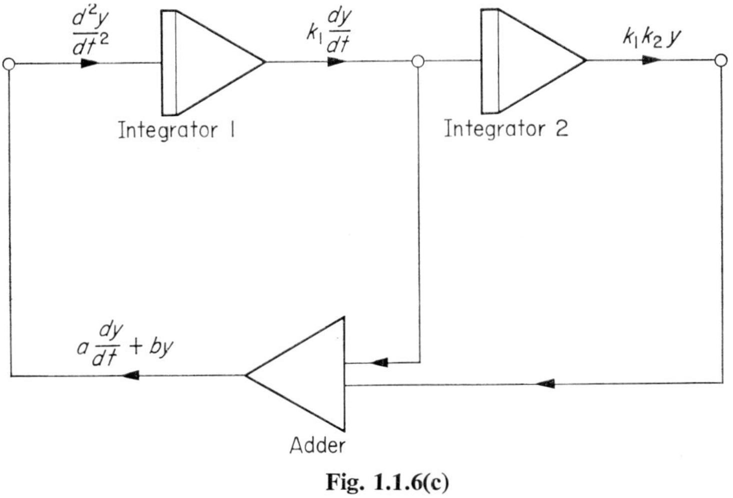

Fig. 1.1.6(c)

on to become the input to the first integrator. Hence when the system is started with pre-set values of $\frac{dy}{dt}$ and y at $t = 0$ the variable y must satisfy the differential equation $\frac{d^2y}{dt^2} = a\frac{dy}{dt} + b$, in which the values of the constants a and b can be chosen at will, in sign, as well as in magnitude.

This is a very simple example which serves to illustrate the methods used, and clearly with more integrators brought into action, quite complicated equations can be solved. The solutions obtained show the continuous variations in the values of the variable in question and can be displayed on a cathode ray tube or drawn on paper in the form of a graph.

The accuracy obtainable from Analogue Computers depends entirely on the accuracy and precision of individual components, and of course when using several units the small errors which occur in each of the units tend to be cumulative.

Good electronic analogue computers are manufactured and widely used in industry, and can give results correct to four significant figures.

1.1.7 The digital computer

A digital computer is built up from a large number of simple elements or cells. These cells are arranged so that they can be used for arithmetic in the binary scale. (*See footnote) and also for logical problems.

In order to use the machine two things must be fed into it:

(a) the data to be operated on, and
(b) detailed instructions of the operations to be performed on the data.

There are two chief ways in which this can be done; some computers are operated by sets of punched cards and others are operated by punched paper tape. In either of these cases the pattern of holes punched according to a code has to be recognised, deciphered, and acted upon by the machine.

It must be realised that although capable of performing certain processes the machine can only do these operations as instructed by the programme it is given.

* 315·4 normally (i.e. in scale 10) means

$$3 \times 10^2 + 1 \times 10 + 5 + \frac{4}{10}$$

but 315·4 (scale 7) means

$$3 \times 7^2 + 1 \times 7 + 5 + \frac{4}{7}$$

and 10101 (scale 2) means

$$2^4 + 0 + 2^2 + 0 + 1$$

i.e. 10101 (scale 2) is the same as 21 (scale 10).

The machine is constructed or can be programmed so that it performs its own conversions from decimal to binary scale and back to decimal again.

The digital computer, basically, must have:

(i) an input system
(ii) a store for numerical data
(iii) a store for instructions
(iv) a control unit
(v) an arithmetic unit
(vi) an output system.

It is only capable of executing certain elementary operations, e.g. take the numbers from two given locations in the store, add them, and store their sum in a third location. Any calculation must therefore be broken down into an ordered sequence of correspondingly simple steps before it can be accepted by the computer. This process usually requires the formulation of a flow chart (like those used in various chapters below) which arranges all the major steps of the problem in logical sequence. Each of these steps must then be further subdivided into the required sequence of the elementary operations in the correct order. This list of instructions is called a Programme, and must now be punched (on cards or tape) according to the code of the particular computer to be used. The task of such detailed machine code programming can be made less laborious by establishing a library of subroutines, i.e. programmes for the operations which are most frequently required (e.g. finding of square-roots). These sub-routines are then available as ready made sections to be incorporated in any main programme as required. Again, these sub-routines can be permanently stored in the computer and called into use when desired by means of an 'Autocode'. Carrying this idea of speeding up and easing the work of programming a stage further we have the computer languages (such as Algol and Fortran). This method of programming involves writing the instructions in a standardised and precise form using code words or symbols each one of which instructs the machine to use one of the many routines already available in its store.

The speed of the digital computer is such that the time taken by each elementary operation is measured in microseconds (millionths of a second). In order to take advantage of this speed, successive operations must be made to follow each other rapidly and therefore automatically. Hence although each operation is extremely elementary, many thousands of them are performed during each working second and we have a very powerful and flexible machine. At first sight, the effect of this speed may seem to be one of degree only, but in practice it makes a very significant difference. This type of machine can handle problems of far greater complexity than one would ever think of giving to a team of experts. Also, the automatic working of the machine makes available the facility of modifying the course of a calculation as it is going on, i.e. of choosing which of several steps to take next in accordance with the way some previous step worked out.

The total time involved in using a computer on a problem is the time required to design the programme together with the time required to prepare the tape (or cards), and test and maybe correct it, and also the actual computer running time. In addition, considerable time may be required to print out the results from the output.

The speed of operation and the method of programming combine to make the digital computer an ideal machine for repetitive processes such as are used in successive approximation methods (See Chapter 4). For example it is so good at finding square roots by Newton's method $\frac{1}{2}(x_r + N/x_r)$ (§ 4.2 end) that it is quicker to start the machine off with x_0 equal to a large number than to specially obtain and feed in a reasonably close first approximation.

The accuracy obtainable when using a digital computer is independent of the accuracy of the individual components, and is limited only by the number of digits which it is possible to programme and accommodate.

1.2 USEFUL TECHNIQUES FOR ANALYSING THE NATURE OF FUNCTIONS

1.2.1. The binominal theorem is most useful in the form:

$$(1 + x)^n = 1 + \binom{n}{1}x + \binom{n}{2}x^2 + \dots\dots + \binom{n}{r}x^n + \dots\dots$$

where
$$\binom{n}{r} \equiv \frac{n(n-1)(n-2)\dots(n+1-r)}{1.\,2.\,3.\dots r}$$

(a) if n is a positive integer the expansion of $(1 + x)^n$ terminates at the term x^n e.g. $(1 + x)^5 \equiv 1 + 5x + 10x^2 + 10x^3 + 5x^4 + x^5$.

(b) if n is not a positive integer the expansion becomes an infinite series and is valid only if $x^2 < 1$.

e.g. (i) $\sqrt{(1 + x)} = (1 + x)^{\frac{1}{2}} = 1 + (x/2) - (x^2/8) + (x^3/16) - (5x^4/128) + \dots$ to ∞ (provided $x^2 < 1$)

(ii) if $x^2 > 1$ we could use $\sqrt{(1 + x)} = \sqrt{[x(1 + 1/x)]}$ and obtain $\sqrt{(1 + x)} = \sqrt{x}[1 + 1/x]^{\frac{1}{2}} = \sqrt{x}[1 + (1/2x) - (1/8x^2) + (1/16x^3)] + \dots$

(iii) $(a + b)^n = a^n[1 + (b/a)]^n$ (useful when $b < a$)
or $= b^n[1 + (a/b)]^n$ (useful when $b > a$)

(c) The first few terms of the series can be used as an approximation provided the sum of the remainder of the terms is less than the acceptable error. In many cases if the successive terms are decreasing rapidly enough in value the first discarded term can be used as an estimate of the value of this truncation error.

e.g. (i) $\sqrt{1{\cdot}001} = 1 + \frac{1}{2}(0{\cdot}001) = 1{\cdot}0005$ provided that $\frac{1}{8}(0{\cdot}001)^2 \simeq 0{\cdot}000\,000\,1$ can be ignored,

i.e. provided that not more than six decimal places are required.

e.g. (ii) If $x = 0{\cdot}123$, $\sqrt{(1 + x)} \simeq 1 + (x/2) - (x^2/8) + (x^3/16)$ provided that the next term $\dfrac{5x^4}{128} = \dfrac{5 \times 0{\cdot}000\,22\dots}{128} < 0{\cdot}000\,01$ can be ignored.

This approximation would therefore be correct to 4D (i.e. correct to four decimal places).

1.2.2 Taylor's theorem

The expansion of $f(a + h)$ in ascending powers of h is known as Taylor's theorem:

$$f(a+h) = f(a) + hf'(a) + \ldots + (h^n/n!) f^n(a) + \ldots$$

where $f^n(a)$ denotes the value of $(d^n/dx^n)[f(x)]$ at $x = a$, and can be used whenever $f(a)$, $f'(a) \ldots f^n(a) \ldots$ exist and are finite, and is valid for the range of values of h for which the resulting series is convergent.

e.g. if $f(x) \equiv \log x$ then $f'(x) = (1/x), f''(x) = -(1/x^2), f'''(x) = (2/x^3) \ldots$

$$\therefore \log(a+h) = \log a + (h/a) - (h^2/2!).(1/a^2) + (h/3!).(2/a^3) \ldots$$

$$= \log a + (h/a) - (h^2/2a^2) + (h^3/3a^3) - \ldots$$

and is valid when $a > 0$, $a + h > 0$, and $-1 < (h/a) \leqslant +1$.

If $a = 1$ we have $\log(1 + h) = h - \frac{1}{2}h^2 + \frac{1}{3}h^3 - \ldots$

which is valid when $-1 < h \leqslant +1$.

Example: Expand $\cos(a + h)$ and hence verify that

$$\cos h = 1 - (h^2/2!) + (h^4/4!) + \ldots$$

Taylor's series is frequently used to analyse the nature of a function $f(x)$ in the neighbourhood of the point $x = a$ in cases where this point presents certain difficulties or is especially interesting.

In simple cases, especially if only the first few terms are required, the expansion can often most easily be obtained by using standard expansions (see 3.4. examples a, b, c). e.g. Consider $y = (1 - \sin x)/(\pi - 2x)$ near $x = (\pi/2)$. Putting $x = (\pi/2) + h$ we have:

$$y = \frac{1 - \cos h}{-2h} = \frac{1 - [1 - (h^2/2!) + (h^4/4!) - \ldots)}{-2h}$$

$$= -(h/4) + (h^3/48) - \ldots$$

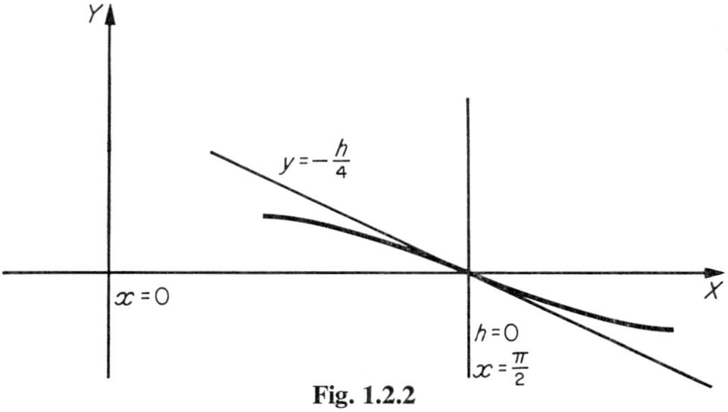

Fig. 1.2.2

showing that when $h = 0$ [i.e. $x = (\pi/2)$] we have $y = 0$, and that y is approximately equal to $-(h/4)$ for small h so that plotting y against h gives a straight line through $h = 0$, $y = 0$ with gradient $-\frac{1}{4}$. If we now include the term in h^3 this brings values of y above the straight line when h is positive and below when h is negative. Hence we have the preceeding (Fig. 1.2.2) sketch for small values of h, i.e. for values of x near $x = (\pi/2)$ [N.B. $x = (\pi/2) + h$]. [See also 3.1.1.]

The graph of $y = (1 - \sin x)/(\pi - 2x)$ therefore passes through the point $[(\pi/2), 0]$ with gradient $-\frac{1}{2}$ and has a point of inflexion there.

1.2.3 Maximum and minimum values of a function

If $V \equiv f(x)$, a function of one variable only, then the equation $(dV/dx) = 0$ gives all the values of x at which V may have stationary values. Further tests must be applied to determine whether V is a maximum, or a minimum, or neither at each of these values of x.

If $V \equiv f(x, y)$, a function of two independent variables, then we use the symbol $(\partial V/\partial x)$ to represent the rate of change of V with respect to x when only x is changing, i.e. y is treated as a constant while $(\partial V/\partial x)$ is obtained by differentiating $f(x, y)$ with respect to x. Similarly $(\partial V/\partial y)$ is obtained if x is regarded as a constant while $f(x, y)$ is differentiated with respect to y. The two simultaneous equations $(\partial V/\partial x) = 0$ and $(\partial V/\partial y) = 0$ give all the values of x and y (in the form $(x_1\ y_1)$, $(x_2\ y_2)$ . . . etc.) at which V may possibly have stationary values. Further tests must be applied to determine whether V actually is a maximum or a minimum at any of these values of x and y.

In many cases the testing for actual maximum or minimum points is simplified by the physical nature of the problem.

Example 1

A right pyramid of height h has a square base of side a and its total surface areas is fixed at S. Find, in terms of S, the values of a and h for which its volume (V) is a maximum.

It is clear from the start of this problem that as h increases from zero, a decreases from $\frac{1}{2}\sqrt{(2S)}$, and consequently $V (= \frac{1}{3}a^2h)$ will at first increase, rise to a maximum and subsequently decrease. Hence when we show that $36V^2 = a^2 S^2 - 2a^4 S$ and that $(\partial V/\partial a) = 0$ when $a = \frac{1}{2}\sqrt{S}$ we know that this *is* the value of a which makes V a maximum (within the range of values for which the problem is a real one).

Example 2

Find the co-ordinates (a, b) of the point P such that the sum (S) of the squares of its distances from the fixed points $(x_1\ y_1)$, (x_2, y_2) . . . (x_r, y_r) . . . (x_n, y_n) may be a minimum.

We have $S = \sum_{r=1}^{n} \{(a - x_r)^2 + (b - y_r)^2\}$

$$\therefore (\partial S/\partial a) = \sum_{r=1}^{n} 2(a - x_r) = 2(na - \Sigma x_r)$$

$$\text{and } (\partial S/\partial b) = \sum_{r=1}^{n} 2(b - y_r) = 2(nb - \Sigma y_r)$$

Hence $(\partial S/\partial a) = 0$ and $(\partial S/\partial b) = 0$ when $a = (\Sigma x_r/n)$ and $b = (\Sigma y_r/n)$ and from the nature of the problem we can be certain without further test that these values of a and b will make S a minimum.

1.3 HAND-CALCULATING MACHINES

We assume that a student undertaking a course of elementary Numerical Analysis will have available the most important aid for this type of work, namely a desk hand-calculating machine. In the remainder of this chapter we aim to enable the student to master the elementary arithmetical processes on the machine. The processes involved will be used continuously in the later chapters and then the student must be able to select the most suitable method to use for any particular problem, which is not only the most precise and suitable but also leads to an answer with the required degree of accuracy.

Most of the calculations in this book have been done using a machine which incorporates in its mechanism an ability to 'Back Transfer' any number in the Accumulator on to the keyboard and setting register automatically.

We shall refer to the main parts of the machine in the following ways:

(i) *Setting Keyboard*—Using either 'setting levers' or pressing keys to set a number.

(ii) *Setting or Checking Register* (S.R.) used to check that the keyboard setting is correct.

(iii) *The Accumulator or Product Register* (Acc.) The register which records the results of sequences of additions, subtractions or multiplications.

(iv) *The Counting or Multiplying Register* (C.R.) This records the equivalent number of turns performed by the crank handle.

(v) The Operating Handle or Crank.

1.4 PRESENTATION AND ARRANGEMENT OF WORK

In numerical work, a neat and tidy presentation of the work is essential. Care with recording helps in reducing the possibility of mistakes and renders the location and correction of errors a simple task. Results should be set out systematically, showing all the necessary intermediate steps in the working. Ruled paper or squared paper is very helpful in this respect.

Sufficient information concerning the question and the method of solution should always be given with the original solution.

1.5 MISTAKES AND ERRORS

We now consider more fully some of the points first mentioned.

Checks should be continually introduced as integral parts of the numerical processes being used. In this way, the possibility of mistakes occuring, whether human or inherent to the calculating processes involved, are reduced to a minimum.

1.5.1 Sources of human error

(i) Mistakes occur through carelessness with decimal points and algebraic signs.

(ii) Transposition of digits from a register of the machine onto paper, e.g. writing 3176 as 3716.

(iii) Wrongly transcribing a number which involves repeated digits, e.g. writing 135 516 as 133 516.

(iv) Inaccurate use of tables.

1.5.2 Other types of error are often unavoidable since they are a result of the formulae or processes being used. Such errors may be due to the following reasons.

(i) The approximate nature of the formula used, e.g. if only a finite number of terms of an infinite series are used in an evaluation. The error involved in ignoring all the terms after the finite number which are retained is called '*truncation error*' (Two of the examples in Vol. II. § 1.4 demonstrate the use of a truncated formula.)

(ii) If only the more significant digits of a number are retained then we automatically introduce, what is called a '*rounding-off error*'. Any further calculation, involving such a quantity, will also contain an error. (For further consideration see §1.11 and §5.1.3.)

Thus in all numerical work it is essential that the student should introduce, wherever possible, adequate checks, either mental or numerical, to guard against the mistakes and errors mentioned above. You will notice that in many of the examples in this book, checks are suggested and used.

Remember, no human being or machine is infallible, all numerical work requires great care and checks should be regarded as a safeguard against the occasional mistake which may occur, even in careful work.

Only in this chapter do we explain fully the machines operations, attempting to make clear how the final result, in most cases, is obtained by a 'repeated sequence of operations'. The operator should at all times consider such sequences to see if they are the best available, thus leading to the result in the quickest time and to the required degree of accuracy.

1.6 ADDITION

To add $365 + 52 + 1231$.

Method:

(i) Clear all registers and set the carriage as far to the left as possible. (From now on we shall refer to this as the Normal Starting Procedure.)
(ii) Set 365 on the extreme right of the keyboard. Check via setting register S.R.
(iii) Rotate the crank once in a clockwise direction and the 365 appears on the Accumulator Register. (Acc.)

(iv) Clear keyboard.
(v) Set 52 on extreme right of the keyboard. (Check setting on the S.R.)
(vi) Crank clockwise for addition.

(vii) Clear keyboard.
(viii) Set 1231 on the right of the keyboard. (Check on S.R.)
(ix) Rotate crank for addition.

Now the accumulator register shows the required sum with a 3 in the counting register (C.R.).

C.R.									Acc.												
0	0	0	0	0	0	0	3		0	0	0	0	0	0	0	0	0	1	6	4	8

(This picture refers to a machine possessing an Accumulator with a capacity of 13 figures and a C.R. capacity of 8 figures.)

The 3 in the C.R. shows that three operations of the crank have been performed and that the three numbers are included in the final accumulated total of 1648.

Notice how a sequence of three operations is repeated three times in the above addition sum. In using an electronic digital computer instructions such as these are 'programmed' i.e. the instructions are broken down into the most simple steps and arranged in logical order. One technique used by computer programmers is that of a *FLOW CHART*, which will show clearly the sequence of operations in this logical order for the method used in solving the problem.

1.6.1* A flow-chart

In the foregoing example we have bracketed together the sequence of three operations which is repeated three times before the calculation is completed. We shall continue to use this method of demonstrating repeated sequences.

* For a fuller discussion of Flow Charts see Chap. 2.

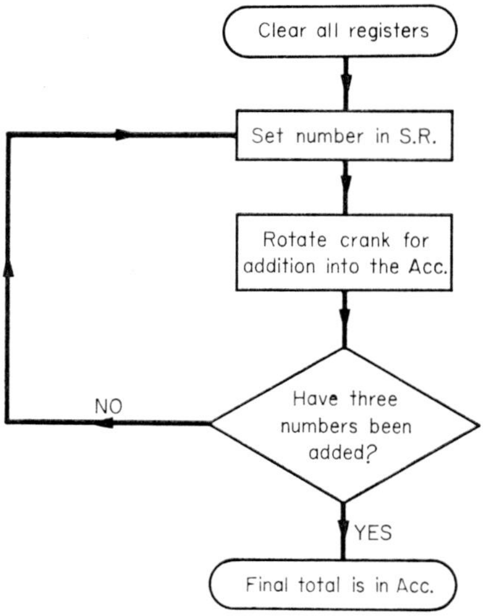

Fig. 1.6.1. A flow chart showing the addition of three numbers using a hand-calculating machine.

1.7 SUBTRACTION

Subtraction is achieved by rotating the crank in an *anti*-clockwise direction. Consider 1648 − 969

Method:

(i) The Normal Starting Procedure (as stated at start of §1.6).
(ii) Set 1648 on extreme right of keyboard and check S.R.
(iii) Rotate crank clockwise.
(iv) Clear keyboard.
(v) Set 969 on extreme right of the keyboard. Check S.R.
(vi) Rotate crank for subtraction.

This gives

			C.R.												Acc.						
0	0	0	0	0	0	0	0		0	0	0	0	0	0	0	0	0	0	6	7	9

Note that the C.R. now shows all noughts, can you explain this?

1.8 DECIMALS ON THE MACHINE

Having considered the basic operations of addition and subtraction with whole numbers we now consider the operations with decimal quantities. Many errors which occur when one first begins using a hand-calculating machine are caused through the careless manipulation of the decimal point markers.

Consider

$$16{\cdot}16 + 15{\cdot}017 + 315{\cdot}2$$

We must now use the decimal-point markers on each register.

By inspection we see that the answer will contain three figures after the decimal point. Hence we set the decimal point markers on the S.R. and Acc. to the necessary positions between the third and fourth spaces.

N.B. Whenever decimal numbers are involved the use of the decimal point markers is essential.

Some people advocate the use of the markers, even when decimal quantities are not involved. They suggest that, if the markers are set in multiples of three spaces along each register, one is able to recognize more readily the magnitude of a number and so minimise the possibility of an error occurring when noting down the number.

For the above example,

Method:

(i) Normal Starting Procedure and set the decimal-point markers.
(ii) Set 16·16 in correct position on the right of the keyboard, i.e. the S.R. should read 16·160
(iii) Rotate the crank for addition.

(iv) Clear keyboard.
(v) Set 15·017 on keyboard. Check S.R.
(vi) Crank for addition.

(vii) Clear keyboard.
(viii) Set 315·2 on keyboard. Check S.R. 315·200.
(ix) Crank for the final addition.

The Acc. then shows the required total.

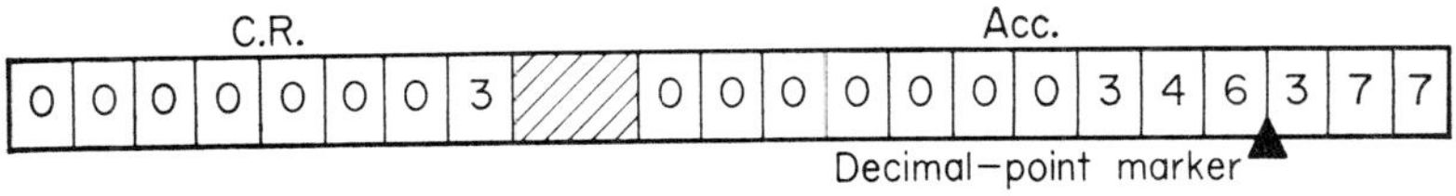

Again the above problem is an example in which a sequence of operations has been repeated.

By now you will realise that the machine can only perform the basic operations of addition and subtraction. Every other calculation to be performed must be so 'programmed' that only addition and subtraction are involved.

1.8.1 An example to consider the sequence of operations in

$$15{\cdot}12 + 371{\cdot}063 - 169{\cdot}19$$

Method:

(i) N.S. Procedure and set the decimal-point markers between the third and fourth spaces.
(ii) Set 15·120 on keyboard and check S.R.
(iii) Crank clockwise.

(iv) Clear keyboard.
(v) Set 371·063 on keyboard. Check S.R.
(vi) Crank clockwise.
(vii) Clear keyboard.
(viii) Set 169·190 on keyboard. Check S.R.
(ix) Subtract by rotating crank anti-clockwise.

The answer shown in the Acc. is 216·993.

1.9 NEGATIVE NUMBERS

We now consider one of the most interesting properties of a hand-calculating machine. We shall observe how the machine signals that the quantity in the accumulator register has changed from positive to negative.

To look more closely at this, perform on the machine the subtraction

$$15 - 24$$

On performing this subtraction the machine rings a bell and at the same time a 'signal' runs across the Acc. showing itself as a series of nines. If this subtraction is performed slowly enough the operator will see that the nines do start at the right hand side of the Acc. and run across the register one at a time. This series of nines is sometimes called '*The Bridge of Nines*'. The bell and the nines signal that the quantity in the Acc. has changed sign.

The Acc. now shows

C.R.									Acc.												
0	0	0	0	0	0	0	0		9	9	9	9	9	9	9	9	9	9	9	9	1

This number showing in the Acc. is the *complement* of 9 or the *complementary form* of the negative number −9. It can be regarded that this is the machine's method of representing the directed number −9. In this 'complementary form' only the figures to the right of the 'Bridge of Nines' are the significant figures.

It should be noted that a number and its complement will always add up to a total, determined by the capacity of the machine being used. In the above machine in which the Acc. has provision for thirteen digits, this total is 10 000 000 000 000, although the figure 1 will not be shown on the machine.

The translation from the 'complementary form' to the modulus of the negative number, viz. $|-9| \Rightarrow 9$ onto the Acc. is achieved, using the back-transfer ability of the machine as follows:

(i) Back transfer the number in the Acc. on to the keyboard.
(ii) Then subtract this number from the clear accumulator (i.e. from 10^{13}).

The Acc. now shows a nine on the extreme right with three nines on the extreme left.

C.R.									Acc.												
0	0	0	0	0	0	0	1		9	9	9	0	0	0	0	0	0	0	0	0	9

The three nines on the left of the Acc. are spurious and remain, only because the capacity of the keyboard is 10 digits. In most of the cases we consider, these three nines are easily distinguishable from the significant figures in the modulus of the negative number.

1.9.1 Mental check. It is not a difficult matter to perform a mental check when the 'Complementary Form' of the negative number is in the accumulator.

Consider 141 − 1367

Performing this subtraction gives in the Acc.

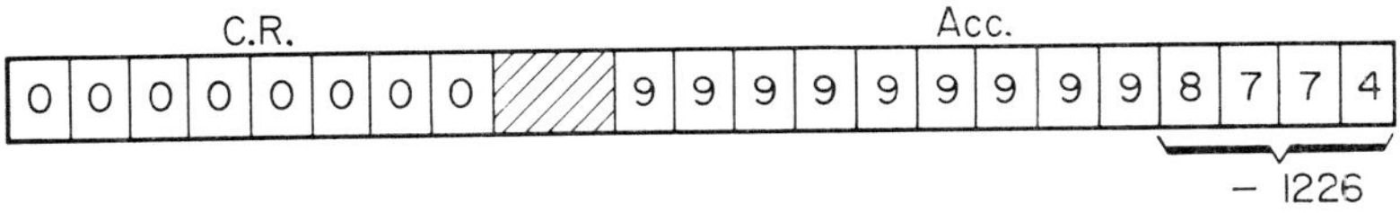

The mental answer is obtained by starting from the extreme right and subtracting the first non-zero digit from 10 (the 4 in this case). Then moving to the left all the remaining numbers are subtracted from nine, to give the answer, as shown above, −1226.

This answer should be checked by performing the further subtraction as described in the previous example.

To further understanding of this machine property, carry out the following example

$$15 - 24 - 10 + 20$$

and perform the operations in the order as given, remembering to crank clockwise for positive numbers and anti-clockwise for subtraction. It is easy to follow how the machine, when performing such a sequence of operations, shows the change from positive to negative numbers and vice-versa, with the usual signals.

1.9.2 An example involving decimals:

$$1{\cdot}361 + 5{\cdot}26 - 7{\cdot}132$$

Method

By inspection, we set the decimal-point markers on S.R. and the Acc. between the third and fourth space.

(i) Normal Starting Procedure.
(ii) Set 1·361 on the keyboard. (Check S.R.)
(iii) Rotate crank for addition.
(iv) Clear keyboard.
(v) Set 5·260 on keyboard. (Check S.R.)
(vi) Rotate crank for addition.

(vii) Clear keyboard.
(viii) Set 7·132 on keyboard. (Check S.R.)
(ix) Rotate crank for subtraction.

The Acc. now shows the 'Complementary Form' of the negative answer.

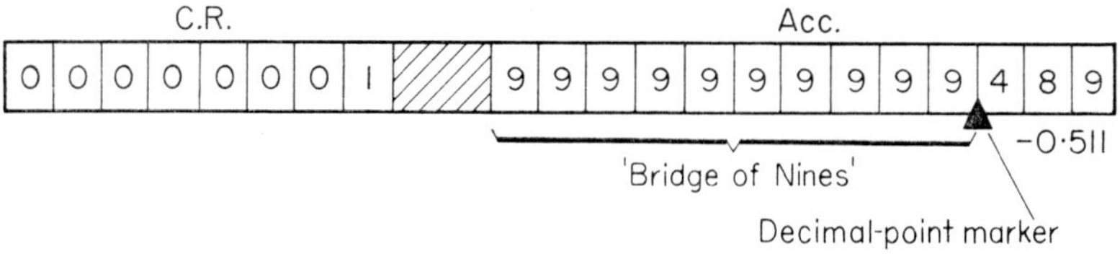

Note that the negative number shown in the Acc. is the complement of $|-0{\cdot}511|$ to give 10^{10}.

Mental evaluation, as well as a check on the machine, gives the answer to the problem as $-0{\cdot}511$.

1.10 EXAMPLES

1. Evaluate $13{\cdot}71 + 7{\cdot}316 - 52{\cdot}278$
2. Evaluate $0{\cdot}316 + 17{\cdot}61 - (23{\cdot}453 - 4{\cdot}137)$
3. Evaluate $-5{\cdot}176 + 10{\cdot}16 - 6{\cdot}31$

(*N.B.* Method: Setting 5·176 on keyboard, crank anti-clockwise because it is a negative number. Set 10·16 and crank clockwise. Set 6·310 and crank anti-clockwise. The 'complementary form' of the negative answer is then shown in the Acc.)

4. Evaluate

$$-15{\cdot}3165 + 9{\cdot}16 + 11{\cdot}1371 - 4{\cdot}9807$$

5. Sum the following array of quantities 'across the rows' and 'down the columns'. We use the final check sum, as a check on the accuracy of the calculations; in which we add the four answers across the bottom and also the three answers on the right hand side of the array, which should give the same answer.

− 1·3275	15·1	21·63	−32·817 →	________
3·12	−11·17	−12·371	20·216 →	________
−41·216	10·8321	23·68	− 1·1553 →	________
↓	↓	↓	↓	FINAL CHECK SUM
________	________	________	________	________

1.11 EFFECT OF 'ROUNDING-OFF' ERRORS

Before proceeding to multiplication and division we consider the effect of rounding-off errors on the degree of accuracy in problems involving addition and subtraction.

Introducing a suffix notation which will be used for the remainder of the chapter, let A_r be a rounded-off value and A_e the corresponding exact value of a quantity, not necessarily known.

Then $$A_r - A_e = a \text{ where '}a\text{' is called the } \textit{absolute error.}$$

thus $$A_e = A_r - a \qquad (1)$$

Then 'a' the absolute error will be positive or negative according as A_r is greater than or less than A_e respectively. So that $|a|$, which is the modulus of 'a' and implies the magnitude only of the absolute error, will be known as the *absolute error modulus of A_r.*

Suppose a set of numerical observations, suitably rounded-off, are represented, in notation, by

$$(A_r)_3 = (A_r)_1 + (A_r)_2 \quad \text{which are approximations for}$$

$$(A_e)_3 = (A_e)_1 + (A_e)_2$$

then using (1) and substituting for $(A_e)_r$ we have

$$(A_r)_3 - a_3 = (A_r)_1 - a_1 + (A_r)_2 - a_2$$

Now cancelling the $(A_r)_3$ on the L.H.S. with the $(A_r)_1 + (A_r)_2$ on the R.H.S. and dividing through by -1 gives

$$a_3 = a_1 + a_2$$

If these three quantities have the same sign, all positive or negative, then the equality is true when the modulus of each quantity is taken, giving

$$|a_3| = |a_1| + |a_2|$$

However, if the three quantities differ in sign, then with a_3 positive and the two terms on the right hand sign differing in sign, we will assume that a_2 is negative. In this case a_1 must be greater than a_3 giving

$$|a_3| = |a_1| - |a_2|$$

from which we conclude that

$$|a_3| < |a_1| + |a_2|$$

and combining these results, we have,

$$|a_3| \leqslant |a_1| + |a_2| \qquad (2)$$

With similar reasoning, if we had $(A_r)_3 = (A_r)_1 - (A_r)_2$ we would obtain the same result (2).

Further this result can clearly be applied to the sum of any number of terms.

In words, this result (2) is that, the absolute error modulus in an addition or subtraction sum is less than or equal to the sum of the absolute error moduli of the separate terms.

1.11.1 *Worked Example*

Consider the numerical example 1·015 + 0·3572 in which both quantities are rounded off. (These two values correspond to $(A_r)_1$ and $(A_r)_2$ above.)

If 1·015 [$(A_r)_1$] has been rounded off to three decimal places then the true value must lie between 1·0145 and 1·0155 so that the absolute error a_1 is such that $-\frac{1}{2} \times 10^{-3} < a_1 < \frac{1}{2} \times 10^{-3}$, which implies that the absolute error modulus is never greater than $\frac{1}{2} \times 10^{-3}$ and this is expressed in the form $| a_1 | \leqslant \frac{1}{2} \times 10^{-3}$.

(N.B. if a number is rounded off to n decimal places then the absolute error modulus $\leqslant \frac{1}{2} \times 10^{-n}$).

Also in the above example the 0·3572 [$(A_r)_2$] gives $| a_2 | \leqslant \frac{1}{2} \times 10^{-4}$

Then

$$\begin{cases} 1{\cdot}015 + 0{\cdot}3572 = 1{\cdot}3722 \\ (A_r)_1 + (A_r)_2 \quad = (A_r)_3 \end{cases}$$

where $| a_1 | \leqslant \frac{1}{2} \times 10^{-3}$ $\qquad | a_2 | \leqslant \frac{1}{2} \times 10^{-4}$

then using (2) $| a_3 | \leqslant | a_1 | + | a_2 |$

gives $| a_3 | \leqslant \frac{1}{2} \times 10^{-3} + \frac{1}{2} \times 10^{-4}$

i.e. $| a_3 | \leqslant (0{\cdot}5 + 0{\cdot}05) \times 10^{-3}$

$| a_3 | \leqslant 0{\cdot}55 \times 10^{-3}$

Hence the exact value of the above addition sum must be between

$$1{\cdot}3722 \pm (0{\cdot}55 \times 10^{-3})$$

i.e. between 1·372 75 and 1·371 65 so that this answer may be *correctly* rounded to 1·37; or if rounded to 1·372 there is a possible error of $\pm$ 1 in the third figure after the decimal point.

The main contribution to the absolute error modulus comes from the term $0{\cdot}5 \times 10^{-3}$, showing that the dominant factor in the accuracy of addition or subtraction is the quantity (or quantities) with the least number of significant figures after the decimal point.

Before completing this section we introduce some elementary standard notation. In the above example where the exact result was between 1·372 75 and 1·371 65 the answer should be given as 1·37 to three significant figures and this is written 1·37 (3S). Alternatively the answer would be given as 1·37 to two decimal places or in standard notation 1·37 (2D).

1.12 MULTIPLICATION

Consider the multiplication of

$$12{\cdot}37 \times 8{\cdot}9$$

Set the decimal point markers for multiplication

The setting register (S.R.) 2 places for 12·37

Counter register (C.R.) 1 for the 8·9

Then under the normal rules of multiplication the accumulator (Acc.) will then require 3 places after the decimal point.

N.B. In general, for multiplication the decimal-point markers are set so that the number of figures after the decimal-point satisfy the following equation

The number in S.R. + the number in C.R. = the number in the Acc. (3)

Also to be noted is the fact that we multiply by the number with the fewer digits.

Further it is good technique, always to begin with the left hand digit of the multiplier, that is the most significant digit of the multiplier, in this case the eight. The reason for this is twofold. Firstly the Acc. will immediately give a close approximation to the answer. Secondly, when the calculation has been completed the machine will be in the normal starting position.

1.12.1 Method A

(i) Normal Starting Procedure including the decimal-point markers as above.
(ii) Set 12·37 on the right of the keyboard.
(iii) To multiply by the 8·9, move the carriage one place to the right and multiply by 8, the more significant figure of the multiplier, by cranking clockwise 8 times.
(iv) Move the carriage one place to the left and crank clockwise 9 times thus setting 8·9 on the C.R. with 17 turns.

This gives

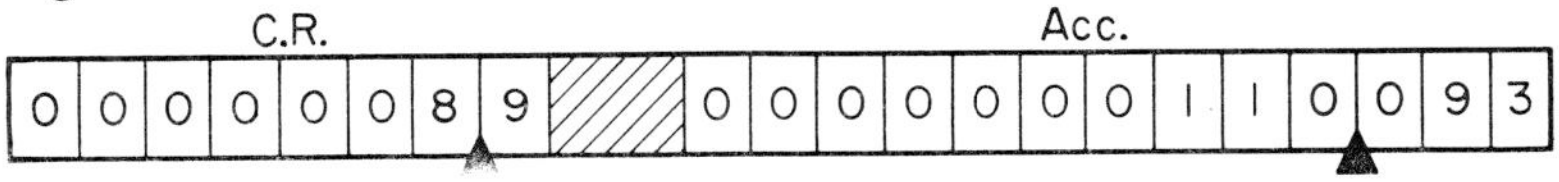

so that 12·37 × 8·9 = 110·093

1.12.2 Quicker Method B

As stated above in the method, it took 17 turns. The calculation can be done more quickly, using fewer turns of the crank. First multiply by 10·0, moving carriage two places to the right and cranking clockwise once. Move carriage one place to the left and crank anti-clockwise once when the C.R. will read 9·0. Finally shift the carriage one place to the left and crank anti-clockwise once. The C.R. will now read 8·9 as required and this has been achieved in three turns of the crank, by means of 10·0 − 1·1.

Consider the multiplication of 157·87 × 17·9 as a further example. The number 157·87 goes on the S.R. Multiply by the 17·9 beginning with one first. If *no* short cuts are taken then the answer will be achieved after 17 turns of the crank.

In the calculation 157·87 × 17·9, using the shorter method we have

$$157{\cdot}87 \times \underbrace{(20{\cdot}0 - 2{\cdot}1)}_{\text{5 turns of crank}} = 2825{\cdot}873 \qquad \text{(takes approx 15 sec.)}$$

1.13 EXAMPLES

1. Evaluate $(12{\cdot}3 \times 8{\cdot}9) + 11{\cdot}61$

2. Evaluate $5{\cdot}381 + (5{\cdot}96 \times 17{\cdot}89)$

Method: by inspection the answer will have four figures after the dec. pt. so (i) Add 5·3810 into the empty Acc. (ii) Clear C.R., (iii) Set dec-pt. markers S.R. 2., C.R. 2. Acc. 4. and complete $17{\cdot}89 \times \underbrace{(10{\cdot}00 - 4{\cdot}04)}_{5{\cdot}96 \text{ in } 9 \text{ turns}}$

3. (a) Evaluate $(3{\cdot}142 \times 16{\cdot}17) \times 3{\cdot}9$

In this example we introduce the *back-transfer property* of the machine, the property by which a number in the Acc. can be transferred to the keyboard automatically. Thus finding the first product and back-transferring this result on to the keyboard enables one to evaluate this triple product as a continuous operation. Care must be taken to set the decimal point markers at each stage.

(b) Evaluate 10! using the back transfer. Set 2, multiply by 3, back transfer, multiply by 4 etc.

4. Evaluate $(15{\cdot}6 \times 7{\cdot}3) + (11{\cdot}31 \times 6{\cdot}1) - (12{\cdot}17 \times 15{\cdot}79)$

By inspection the decimal-point settings are S.R. 2., C.R. 2., Acc. 4. Hence each part of the question needs care with settings e.g. the first product set 15·60 on keyboard and multiply by 7.30. The final product which is preceded by a negative sign, is a negative addition, and so when achieving this product on the machine the crank is turned anti-clockwise.

5. Evaluate $(15{\cdot}61 + 7{\cdot}31) \times 11{\cdot}8$

There are two possible methods of solving this example:

(a) Evaluate the bracket first, back transfer and then multiply by 11·8, or
(b) Evaluate after first removing the brackets giving $(15{\cdot}61 \times 11{\cdot}8) + (7{\cdot}31 \times 11{\cdot}8)$ setting the 11·8 on the keyboard in this case.
(c) Which method is preferable, giving reasons for your answer?

6. Evaluate $(14{\cdot}18 - 16{\cdot}21) \times 21{\cdot}3$

7. Evaluate $4\pi r^2$ where $\pi = 3{\cdot}142$ and $r = 6{\cdot}1$

8. Evaluate $5\pi(r_1^2 - r_2^2)$ where $r_1 = 1{\cdot}32$ and $r_2 = 1{\cdot}24$, $\pi = 3{\cdot}14$

9. In the following addition sum all the quantities are 'rounded-off'

$$1{\cdot}761 + 17{\cdot}32 - 5{\cdot}782$$

Evaluate the absolute error modulus which might be expected and give the answer as accurately as is permitted by the initial data.

1.14 THE EFFECT OF 'ROUNDING-OFF' ERRORS IN MULTIPLICATION

We consider $(A_r)_3 = (A_r)_1 \,.\, (A_r)_2$ which is the product approximating to $(A_e)_3 = (A_e)_1 \,.\, (A_e)_2$
then using equation (1), see §1.11, and substituting we have

$$(A_r)_3 - a_3 = \{(A_r)_1 - a_1\}\{(A_r)_2 - a_2\}$$

giving

$$(A_r)_3 - a_3 = (A_r)_1 \,.\, (A_r)_2 - a_1 \,.\, (A_r)_2 - a_2 \,.\, (A_r)_1 + a_1 \,.\, a_2$$

but we began with $(A_r)_2 = (A_r)_1 \,.\, (A_r)_2$

Which gives $a_3 = a_1 \,.\, (A_r)_2 + a_2 \,.\, (A_r)_1 - a_1 \,.\, a_2$

We now divide both sides by $(A_r)_3 = (A_r)_1 \,.\, (A_r)_2$

Thus $$\frac{a_3}{(A_r)_3} = \frac{a_1}{(A_r)_1} + \frac{a_2}{(A_r)_2} - \frac{a_1 \,.\, a_2}{(A_r)_1 \,.\, (A_r)_2}$$

The final term, which has as its numerator the product of two very small quantities, is then considered negligible. This corresponds to the similar conclusions when multiplying first order increments in the Calculus.

These new quantities we have now achieved, $a_1/(A_r)_1$ are for practical and theoretical considerations known as the '*relative errors*'.

(Theoretically $a_i/(A_e)_i$ should be the relative error, but the above assumption has little, or no effect, on the practical accuracy of this theory.)

Hence we have

$$\frac{a_3}{(A_r)_3} = \frac{a_1}{(A_r)_1} + \frac{a_2}{(A_r)_2}$$

Remembering that the '*absolute errors*' may be either positive or negative, we have, for multiplication

$$\left|\frac{a_3}{(A_r)_3}\right| \leqslant \left|\frac{a_1}{(A_r)_1}\right| + \left|\frac{a_2}{(A_r)_2}\right| \qquad (3)$$

In words: the relative error modulus of a product is less than or equal to the sum of the relative error moduli of the factors of the product.

1.14.1 Example on the above theory:

Consider the product 1·015 × 0·3573 where both quantities have been rounded off.

By machine 1·015 × 0·3573 = 0·362 659 5

Thus the 'relative error modulus' is

$$\left|\frac{a_3}{0{\cdot}363}\right| \leqslant \frac{\frac{1}{2} \times 10^{-3}}{1{\cdot}015} + \frac{\frac{1}{2} \times 10^{-4}}{0{\cdot}3573}$$

$$\leqslant \frac{1}{2{\cdot}03 \times 10^3} + \frac{1}{0{\cdot}7140 \times 10^4}$$

$$\leqslant 0{\cdot}493 \times 10^{-3} + 0{\cdot}14 \times 10^{-3}$$

$$\left|\frac{a_3}{0{\cdot}363}\right| \leqslant 0{\cdot}633 \times 10^{-3}$$

giving $|\, a_3 \,| \leqslant 0{\cdot}633 \times 0{\cdot}363 \times 10^{-3}$

$|\, a_3 \,| \leqslant 0{\cdot}23 \times 10^{-3}$ (actually $0{\cdot}229\,779 \times 10^{-3}$)

Hence the exact value of the product lies between

$$0{\cdot}362\,659\,5 \pm 0{\cdot}000\,23$$

i.e. between 0·362 889 5 and 0·362 429 5

Therefore correctly rounded-off the answer will be 0·36 (2D). If it is rounded-off to 0·363 (3D) we have a possible error of 1 unit in the third figure after the decimal point.

By inspection of the above example, we see that the factor with the larger 'relative error' is the 1·015 and it is this factor which controls the accuracy of the answer. As with addition and subtraction, the factor with the least number of significant figures after the decimal points generally contributes most to the possible error and hence controls the final accepted accuracy.

We leave further examples of accuracy, until after the next section because the calculations involved in using (3) are mainly division.

1.15 DIVISION

We first consider division of whole numbers, using the normal definitions: Dividend is the number to be divided by the Divisor to give the answer, known as the Quotient, with a possible Remainder, if the division cannot be achieved exactly.

We give three possible methods of division, using a hand-calculating machine.

1.15.1 Method A

Consider 737 206 ÷ 361

This method, sometimes called the '*tear-down*' method, corresponds identically to that used when only pencil and paper are available.

(i) Normal Starting Procedure with the carriage in the extreme left position.
(ii) Set 737 206 on the right of the Acc.
(iii) Clear S.R. and C.R.
(iv) Set 361 on the extreme right of the keyboard.
(v) Move carriage the three places to the right until the first three figures (737) of the dividend are immediately beneath the divisor on S.R.
(vi) Subtract repeatedly until the 'Bridge of Nines' appears and the bell rings.
(vii) Crank clockwise until the 'Bridge of Nines' just disappears.
(viii) Move carriage one place to the left.

At this stage compare the pencil/paper method with what is now showing on the machine

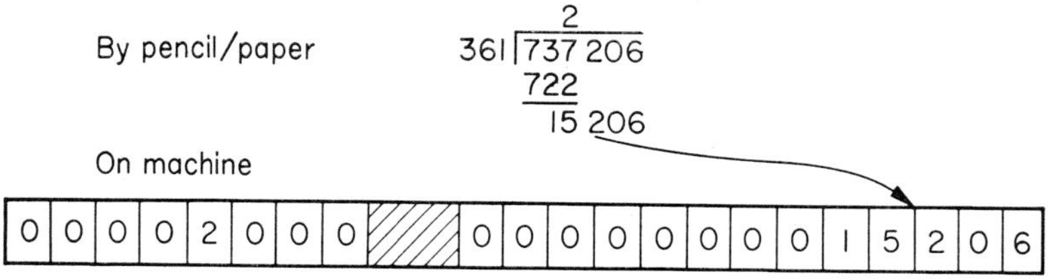

Now repeat the sequence $\left\{\begin{array}{l}\text{(vi)}\\ \text{(vii)}\\ \text{(viii)}\end{array}\right.$ until when repeated, the carriage is the units position.

The machine then shows

C.R. Acc.

0	0	0	0	2	0	4	2		0	0	0	0	0	0	0	0	0	0	0	4	4

giving the Quotient as 2042 with remainder 44.

1.15.2 Method B

Consider again 737 206 ÷ 361

This method is very similar to Method A, but instead of setting the Dividend on the extreme right of the Acc. we begin with the carriage in the extreme right position.

(i) Normal starting procedure and move the carriage to the extreme right.

(ii) Set the dividend on the keyboard, starting in such a position that when transferred to the Acc. it will appear on the extreme left of the Acc.

(iii) Transfer dividend to the Acc.

(iv) Clear S.R. and C.R.

(v) Set 361 on the keyboard, in such a position that it will be subtracted immediately from the 737 of the dividend.

(vi) Subtract 361 repeatedly until the bell rings.

(vii) Crank clockwise until the bell just rings again.

(viii) Move carriage one place to the left.

The Machine now shows

2	0	0	0	0	0	0	0		0	1	5	2	0	6	0	0	0	0	0	0	0

(Compare this with the similar stage in method A).

Repeat $\left\{\begin{array}{l}\text{(vi)}\\ \text{(vii)}\\ \text{(viii)}\end{array}\right.$ a further three times when the Acc. will now read

2	0	4	2	0	0	0	0		0	0	0	0	4	4	0	0	0	0	0	0	0

Notice that in this position we may continue to move the carriage to the left through a further 4 places. Thus in this example, where the division is not exact, this method enables us to work accurately to three decimal places.

Continuing the process, we finish with

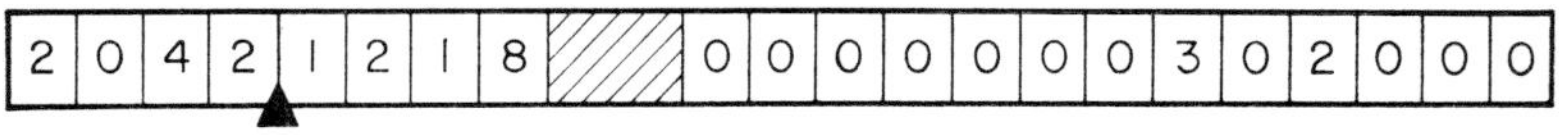

giving an answer 2042·122 (3D)

1.15.3 Method C

The method we are about to consider is known as the 'build-up' method. Consider again 737 206 ÷ 361

(i) Normal Starting Procedure.
(ii) Set 361 on S.R. on the extreme right.
(iii) Move carriage until spaces 6, 5, 4 lie beneath the 361, because the dividend is a six figure number.

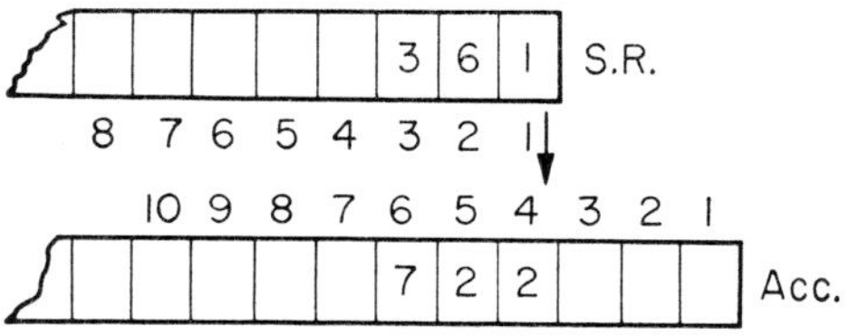

(iv) Add in 361 repeatedly until we get the largest number possible below the required first three digits 737 of the dividend. In this case we get 722.
(v) Move the carriage one place to the left.
(vi) We now add in multiples of 361 until the second figure in the Acc. is a three (in 737 206). But one crank is too much for it *builds-up* the number 758 100, so crank anti-clockwise and retrieve 722 000, as before.
(vii) Move carriage a further place to the left.
(viii) We now add in 361 repeatedly, until the figure in the Acc. is the largest number possible below 7372 . . . which gives 73 644.
(ix) Move carriage a further place to the left.
(x) As before add in 361 repeatedly until we get the largest number possible below the Dividend. This gives 2042 in C.R. and 737 162.

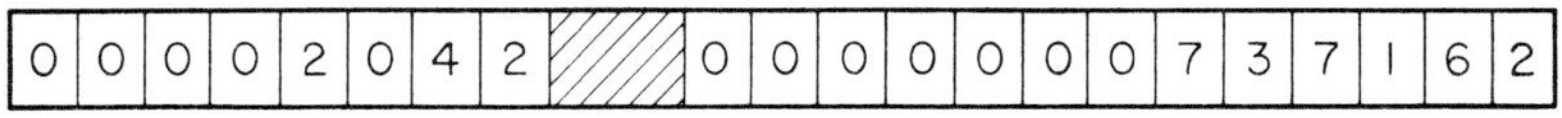

At this stage, we have that

$$316 \times 2042 = 737\,162$$

To find the remainder, if there is one, we set 737 206 in the right of the S.R. and subtract from the final number in the Acc. This will give the 'complementary form' of the actual remainder 44.

This method has drawbacks because

(i) the operator is required to memorise or repeatedly check the dividend.
(ii) When the operation is complete the remainder is NOT automatically shown in the Acc.

Before proceeding to problems involving decimals, work through the examples below using any of the three methods as appropriate, in this way you will not only learn but come to understand the methods.

1.15.4 Examples

1. Evaluate $(315 \times 69) \div 161$
2. $976\,631 \div 97$
3. Evaluate $271 \times \dfrac{1561}{37}$
4. Evaluate $\dfrac{259 \times 754 \times 27}{29 \times 37}$

   Alternative methods:

   a First evaluate and note the value of the denominator. Now multiply out the numerator and divide by the recorded denominator. Thus, apart from the initial calculation and recording, it is continuous sequence of operations on the machine.

   b We may evaluate it as $(259/29) \times (754/37) \times 27$, recording the answers to the first two parts, before performing the multiplication. Since these divisions are not exact, in this example there will be an error in the answer, which should be integral.
   However, in some cases, this could be the best method. For example, if considering, $(a \text{ x } b \text{ x } c)/(d \text{ x } e)$ where a, b, c, are so large that $a \text{ x } b \text{ x } c$ is beyond the capacity of the machine, and similarly for $d \text{ x } e$.

   c Evaluate the numerator and then perform successive divisions.

5. Evaluate $\{(53 \times 719) - (23 \times 79) + 7\} \div 37$

1.15.5 Division involving decimals

We saw in the earlier work how method B automatically gave the opportunity of obtaining the Quotient in decimal form if required.

The method we use here is very similar to that of method A.

Worked Example 1

Consider $17{\cdot}28 \div 2{\cdot}136$

We will work to get three decimal figures in the Quotient. Then a rough check gives $(17{\cdot}28/2{\cdot}136)$ will approximate to 8·xxx

Hence 17·28 = 2·136 × 8·xxx
Acc. S.R. C.R.

Which gives the decimal-point settings as follows:

$$\text{S.R. } (3) + \text{C.R. } (3) \Rightarrow \text{Acc. } (6).$$

Method

(i) Normal Starting Procedure with decimal-point settings as above.
(ii) Set 17·280 000 in Acc.
(ii) Clear S.R. and C.R.
(iv) Set 2·136 on the Right hand side of the S.R.
(v) Move the carriage three places to the right, so that the first figure in the C.R. will appear in the fourth position.
(vi) Continue with the process of division by repeated subtraction as shown in method A of section 1.16.

We finish with

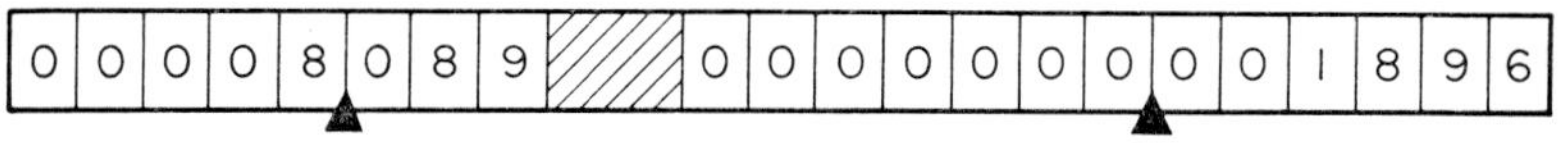

= 8·089

= 8·09 (2D)

Worked Example 2

Consider 72·0532 ÷ 5·172 45

We will work to a quotient of 5 decimal places. The settings therefore, will be S.R. 5., C.R. 5 = Acc. 10.

Repeating the method given above produces an answer

13·9302 (4D) (13·930 18 on C.R.).

1.16 THE EFFECT OF ROUNDING-OFF ERRORS ON DIVISION

Using the same notation as previously.

Suppose

$$(A_r)_3 = \frac{(A_r)_1}{(A_r)_2}$$

is the approximation for the exact division

$$(A_e)_3 = \frac{(A_e)_1}{(A_e)_2}.$$

Then using § 1.11 and substituting in the exact relationship we have

$$(A_r)_3 - a_3 = \frac{(A_r)_1 - a_1}{(A_r)_2 - a_2}$$

$$= \frac{(A_r)_1 \left\{1 - \dfrac{a_1}{(A_r)_1}\right\}}{(A_r)_2 \left\{1 - \dfrac{a_2}{(A_r)_2}\right\}}$$

which is now concerned with relative errors.

Giving

$$(A_r)_3 - a_3 = \frac{(A_r)_1}{(A_r)_2}\left(1 - \frac{a_1}{(A_r)_1}\right)\left(1 - \frac{a_2}{(A_r)_2}\right)^{-1}$$

We now expand the third term on the right hand side using the binomial series and neglect the terms involving products of relative errors, giving

$$(A_r)_3 - a_3 = \frac{(A_r)_1}{(A_r)_2}\left(1 - \frac{a_1}{(A_r)_1}\right)\left(1 + \frac{a_2}{(A_r)_2} + \text{terms of } 2nd \text{ order and above} \ldots\right)$$

$$= \frac{(A_r)_1}{(A_r)_2}\left(1 - \frac{a_1}{(A_r)_1} + \frac{a_2}{(A_r)_2}\right) \text{ omitting the term } \frac{a_1}{(A_r)_1}\cdot\frac{a_2}{(A_r)_2}$$

$$\therefore \quad -a_3 = -\frac{a_1}{(A_r)_2} + \frac{a_2\,(A_r)_1}{\{(A_r)_2\}^2}$$

Then dividing by $(A_r)_3 = [(A_r)_1/(A_r)_2]$ we are able to reintroduce relative errors, to give

$$\frac{a_3}{(A_r)_3} = \frac{a_1}{(A_r)_1} - \frac{a_2}{(A_r)_2}$$

and considering relative error moduli

$$\left|\frac{a_3}{(A_r)_3}\right| \leqslant \left|\frac{a_1}{(A_r)_1}\right| + \left|\frac{a_2}{(A_r)_2}\right| \text{ same as (3)}$$

and by comparison we note that the relative-error modulus for multiplication and division give the same simple formula.

If necessary, we can extend this result to any number of factors in the numerator and denominator. It would simply mean an appropriate increase of terms on the right hand side of (3).

1.16.1 *Example* 1 to show the use of this theory we shall now evaluate the absolute error modulus for the quotient considered in example 1 of section 1.15.5, assuming both quantities are rounded-off.

Using method B for division we obtain

$$\frac{17{\cdot}28}{2{\cdot}136} = 8{\cdot}089\,88 \text{ which corresponds to } \frac{(A_r)_1}{(A_r)_2} = (A_r)_3$$

then substituting in (3), which is

$$\left|\frac{a_3}{(A_r)_3}\right| \leqslant \left|\frac{a_1}{(A_r)_1}\right| + \left|\frac{a_2}{(A_r)_2}\right|$$

We have

$$\left|\frac{a_3}{8{\cdot}09}\right| \leqslant \frac{\frac{1}{2}\times 10^{-2}}{17{\cdot}28} + \frac{\frac{1}{2}\times 10^{-3}}{2{\cdot}136}$$

giving $|\, a_3 \,| \leqslant 0{\cdot}004\,23$

This means that the correct answer to the quotient lies between the values

$$8{\cdot}089\,88 \pm 0{\cdot}004\,23$$

i.e. the correct answer lies in the range

$$8{\cdot}0941 \text{ to } 8{\cdot}0857$$

By inspection of these values, the answer, correctly rounded-off is 8·09 (2D).

If the answer is given as 8·089 then there is a possible error of ± 5 units in the last decimal figure.

1.16.2 *Example* 2, to consider the maximum absolute error in Question 2, §1.13 assuming the given quantities are all rounded-off.

By machine

$$5{\cdot}381 + 5{\cdot}96 \times 17{\cdot}89 = 112{\cdot}0054$$

Considering the multiplication first

$$5{\cdot}96 \times 17{\cdot}89 = 106{\cdot}6244$$

then $\left|\dfrac{a_3}{106{\cdot}6}\right| \leqslant \dfrac{\frac{1}{2} \times 10^2}{5{\cdot}96} + \dfrac{\frac{1}{2} \times 10^2}{17{\cdot}89}$

leading to $|\, a_3 \,| \leqslant 0{\cdot}1194$

i.e. The absolute error modulus for the multiplication is 0·1194.

Therefore, for the whole question, the absolute error modulus

$$|\, a_4 \,| \leqslant \tfrac{1}{2} \times 10^{-3} + 0{\cdot}1194$$
$$\leqslant 0{\cdot}1199$$

so that the correct value of this question lies within the range $112{\cdot}0054 \pm 0{\cdot}1199$ i.e. lies between

$$112{\cdot}1253 \text{ and } 111{\cdot}8855$$

so that correctly rounded-off the answer is 112 (3S).

If the answer is given as 112·0 there is a possible error of ± 1 in the fourth figure.

1.17 SQUARE ROOTS

This method which we are going to discuss is sometimes referred to as the '1, 3, 5,' . . . method of extracting square roots. Later, we shall consider an entirely different approach, namely an iterative method of evaluating square roots.

Basically this method depends upon the fact that the continuous sum of consecutive odd numbers gives the squares of the natural numbers, in ascending order.

For

$$\begin{aligned} &1 \ldots &&= 1^2 \\ &1 + 3 \ldots &&= 2^2 \\ &1 + 3 + 5 \ldots &&= 3^2 \\ &1 + 3 + 5 + 7 \ldots &&= 4^2 \\ &\cdots\cdots\cdots\cdots && \\ &\cdots\cdots\cdots\cdots\cdots\cdots && \\ &1 + 3 + 5 + 7 + 9 + \ldots + (2n - 1) &&= n^2 \end{aligned}$$

(This is easily verified by Induction)

1.17.1 The square root of rounded-off numbers

Worked Example

$\sqrt{(8{\cdot}675)}$.

Using the theory of errors intuitively, it is reasonable to expect that the accuracy of the answer will be limited to 3 decimal figures, so we will work to 4 decimal places, thereby keeping a guarding figure.

$$\sqrt{(8{\cdot}675)} \simeq 2{\cdot}\text{xxxx}$$

In this machine method the answer will appear in the C.R. and we will set the decimal-point markers the same on both the S.R. and C.R.

Giving S.R. (4); C.R. (4) ⇒ Acc. (8)

Work through the example below slowly and carefully.

Method:

(i) Normal Starting Procedure with decimal-point markers S.R. (4), C.R. (4) ⇒ Acc. (8).

(ii) Set 8·675 000 00 in Acc.

(iii) Clear C.R. and S.R.

(iv) We move the carriage 4 places to the right so that the first figure in the C.R. will appear in the fifth position, thus agreeing with the original estimate 2·xxxx.

(v) We now subtract consecutively the odd numbers in ascending order, down the fifth column of the S.R., starting with 1. The bell will ring ('Bridge of Nine' appears) on subtracting five, so crank clockwise until the bell just rings again.

(vi) On the S.R. move the lever setting back one place so that it now reads 4·0000.

The carriage now reads:

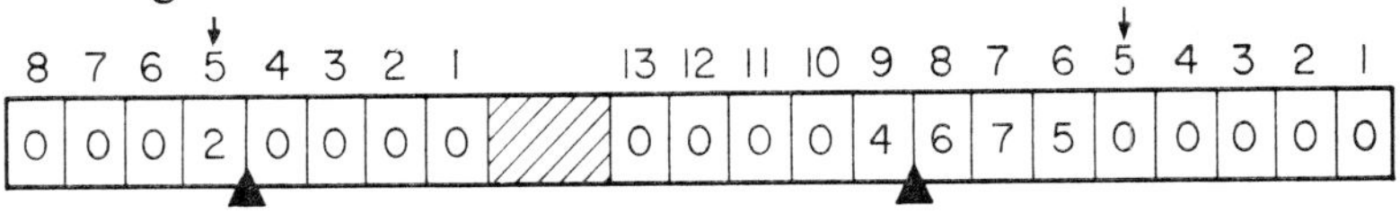

(vii) Move the carriage one place to the left.

(viii) Now subtract successive odd numbers down the 4th column of the S.R. (That is first subtract 41, which the C.R. then tells us is the 21st odd number). Proceeding normally we find that the bell does NOT ring on subtracting 49, so that next we must subtract 51. (Do NOT move to the third column of the S.R.) Continuing the process of subtracting consecutive odd numbers the bell rings when we attempt to subtract 59, so crank clockwise until the bell rings.

(ix) Move fourth lever back one place to give a reading of 58 on S.R.

(Note: The reading on the S.R. is exactly twice the reading on the C.R. and the machine is following exactly the pencil/paper method of extracting square roots). At this stage we have:

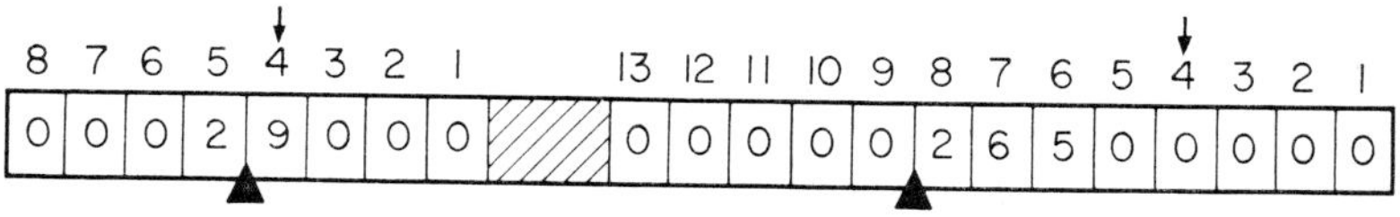

(x) Move the carriage one place to the left.
(xi) Subtract successive odd numbers down the 3rd column of the S.R. Thus the first number to subtract is 581, which the C.R. tells us is the 291st odd number. Continue with method, as described above in (v) and (viii) until the bell rings a second time.
(xii) Move the S.R. back one, when it will read 588.

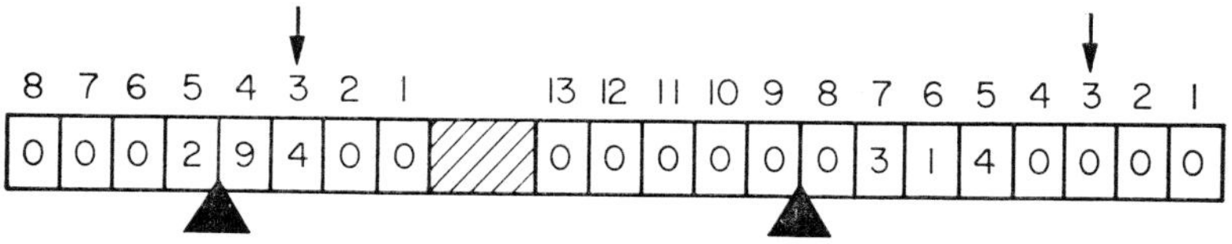

(xiii) Move carriage one place to the left.
(xiv) Subtract successive odd numbers using 2nd lever on S.R. as before.
(xv) Subtract one from 2nd column on S.R.

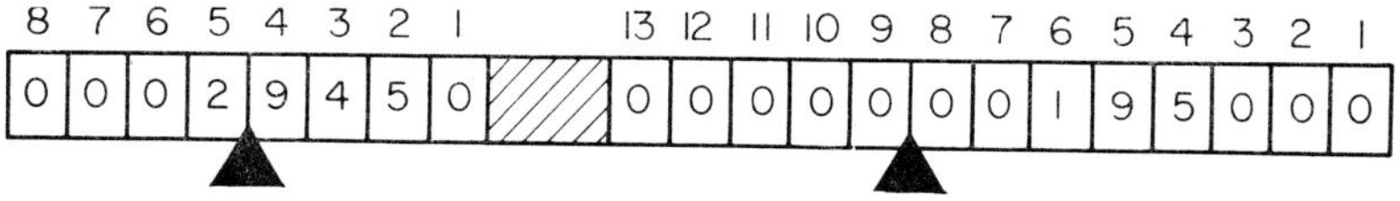

Repeat the sequence of operating for the last time when the answer given in the C.R. is 2·9453.

i.e. $\sqrt{(8{\cdot}675)} = 2{\cdot}9453$

You should always check your result by evaluating, in this case, $(2{\cdot}9453)^2$, which is 8·674 792 09, giving an error of 0·000 207 91. We consider such errors again, in the example below, also in §1.18.2.

At this stage in the chapter, we introduce the notation that will be used in the remainder of the book to demonstrate readings on the machine, at any stage of a computation.

1.17.2 Notation for machine readings

Example

$$\sqrt{(2316{\cdot}674)}$$

Then $\sqrt{('23'16'{\cdot}67'4)}$ will give 4x·xxxx working to 4 decimal places in the C.R. Giving the same setting on the S.R. means C.R. (4); S.R. (4), therefore Acc. 8.

Method	S.R.	C.R.	Acc.
1. Set 2316·674 on Acc. Clear S.R. and C.R.	0	0	2316·674 000 00
2. Move carriage to the right, until in the sixth position. Substract successive odd numbers in 6th column of S.R. starting with 1. Sequence is complete when the bell rings second time after a clockwise crank	90·0000	40·0000	716·674 000 00
3. Subtract one from S.R. in column 6	80·0000	40·0000	716·674 000 00
Move carriage one place to left. Subtract successive odd numbers from 5th column S.R. as 2. above	97·0000	48·0000	12·674 000 00
Subtract one from S.R. in 5th column.	96·0000	48·0000	12·674 000 00
Move carriage one place to left. Subtract successive odd numbers from 4th column in S.R. etc.	96·3000	48·1000	3·064 000 00
Subtract one from 4th column on S.R.	96·2000	48·1000	3·064 000 00
Move carriage one place to left. Subtract successive odd numbers from column of S.R. etc.	96·2700	48·1300	0·177 100 00
Subtract one from 3rd column on S.R.	96·2600	48·1300	0·177 100 00
Move carriage one place to left. Subtract successive odd number from 2nd column of S.R. etc.	96·2630	48·1310	0·080 839 00
Subtract one from S.R. 2nd column	96·2620	48·1310	0·080 839 00
Move carriage one place to left. Subtract successive odd numbers from 1st column of S.R. etc.	96·2637	48·1318	0·003 828 76
Subtract one from 1st column	96·2636	48·1318	0·003 828 76

giving $\sqrt{(2316{\cdot}674)} = 48{\cdot}132$ (3D)

An interesting fact about this method is that if we subtract $(48{\cdot}1318)^2$ from 2316·674 the residual will be 0·003 828 76, as in the last column. (For $(48{\cdot}132)^2$ the residual is 0·015 424).

Further analysis, dealing with errors in powers and roots is given in §1.18.

1.17.3 The flow-chart for this method of finding square-roots is as follows: Consider $\sqrt{(2316{\cdot}674)}$, the example just completed, in § 1.17.2.

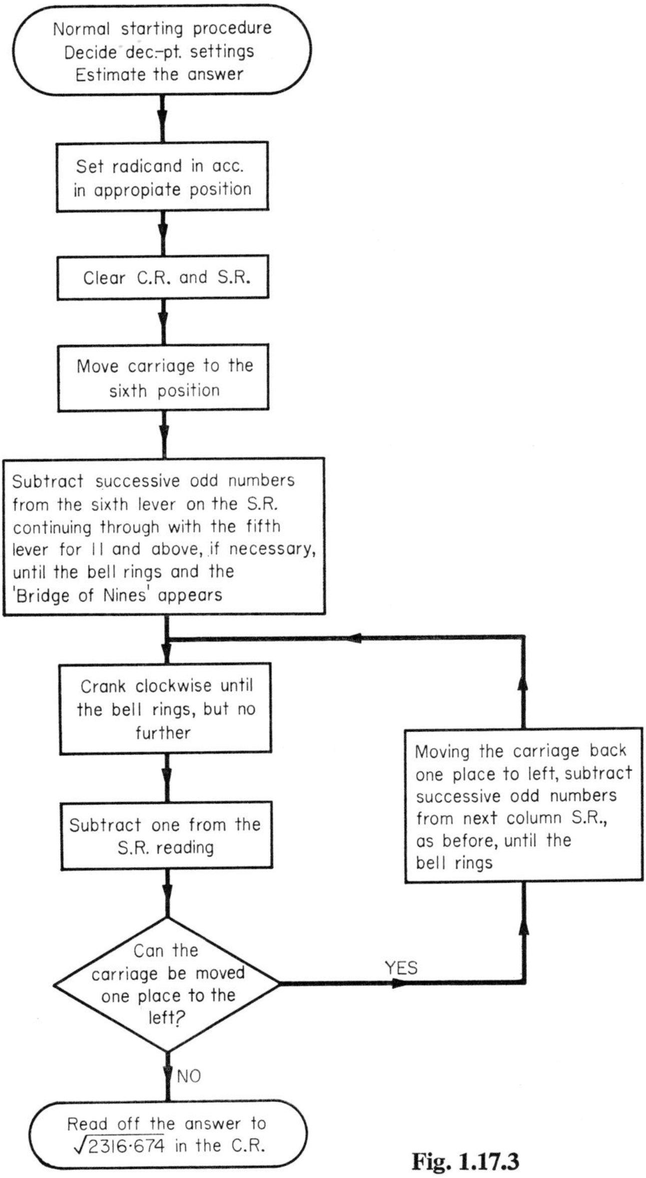

Fig. 1.17.3

1.18 THE EFFECT OF 'ROUNDING-OFF' ERRORS ON POWERS AND ROOTS

Suppose that $(A_r)_2 = \{(A_r)_1\}^p$, where the power 'p' denotes an exact quantity, integral or fractional, is the approximation for the exact relationship $(A_e)_2 = \{(A_e)_1\}^p$.

Then using (1) § 1.11 and substituting in the exact relationship gives

$$(A_r)_2 - a_2 = \{(A_r)_1 - a_1\}^p$$

expanding the right hand side by the Binomial series we have

$$(A_r)_2 - a_2 = \{(A_r)_1\}^p \left\{1 - \frac{a_1}{(A_r)_1}\right\}^p$$

giving $(A_r)_2 - a_2 = \{(A_r)_1\}^p\{1 - p \cdot \dfrac{a_1}{(A_r)_1}$ + terms which are neglected as powers of relative errors}

$$\therefore \quad -a_2 = -\frac{p \cdot a_1\{(A_r)_1\}^p}{(A_r)_1}$$

$$\text{Hence } a_2 = \frac{p \cdot a_1\{(A_r)_1\}^p}{(A_r)_1}$$

Then dividing each side by $(A_r)_2 = \{(A_r)_1\}^p$ respectively we have

$$\frac{a_2}{(A_r)_2} = p \cdot \frac{a_1}{(A_r)_1}$$

giving

$$\left|\frac{a_2}{(A_r)_2}\right| = |p| \cdot \left|\frac{a_1}{(A_r)_1}\right| \qquad (4)$$

In words:

When powers are involved, the relative error modulus of the answer is equal to the product of the modulus of the power and the relative error modulus of the radicand.

1.18.1 Applying this result to the example in § 1.17.1

Considering $\sqrt{(8{\cdot}675)}$, i.e. $(8{\cdot}675)^{\frac{1}{2}}$ which we have found as 2·9453.

Using (4) we have $p = \frac{1}{2}$

so that

$$\left|\frac{a_2}{2{\cdot}945}\right| = \tfrac{1}{2} \times \frac{\frac{1}{2} \times 10^{-3}}{8{\cdot}675}$$

$$= \frac{1}{34{\cdot}7 \times 10^3}$$

$$= 0{\cdot}288 \times 10^{-4}$$

$$|a_2| = 0{\cdot}288 \times 2{\cdot}954 \times 10^{-4}$$

$$|a_2| = 0{\cdot}0001$$

so that the correct value of $\sqrt{(8{\cdot}675)}$ lies in the range

$$2{\cdot}9453 \pm 0{\cdot}0001$$

Hence we may give the result as 2·945 (3D) which is the same degree of accuracy as we concluded intuitively in the previous section.

1.18.2 To show that care must be taken when assuming intuitively the degree of accuracy of an answer, consider the following example.

It would seem reasonable to assume the answer to $\sqrt{(672{\cdot}12)}$ can only be obtained to 2D and as a consequence work to 3 figures after the decimal point.

We have $\sqrt{(672{\cdot}12)} = 2\text{x}{\cdot}\text{xxxx}$

By machine $\sqrt{(672{\cdot}12)} = 25{\cdot}925$

Using (4) $\left|\dfrac{a_2}{25{\cdot}9}\right| = \frac{1}{2}\cdot\dfrac{\frac{1}{2}\times 10^{-2}}{672{\cdot}12}$

giving $|a_2| = 25{\cdot}9 \times 0{\cdot}372 \times 10^{-5}$

$\therefore \quad |a_2| = 0{\cdot}0001$

Giving the correct answer, as lying in the range,

$$25{\cdot}925 \pm 0{\cdot}0001$$

i.e. between 25·9251 and 25·9249
so that the answer, correctly rounded-off is 25·925 (3D) which is contrary to our original assumptions, that only 2D accuracy can be obtained.

Suppose we work to 4 decimal places, by machine we have

$$\sqrt{672{\cdot}12} = 25{\cdot}9252$$

Then, as before, $|a_2| = 0{\cdot}0001$
so that the correctly rounded-off answer lies in the range

$$25{\cdot}9252 \pm 0{\cdot}0001$$

Thus, as before, the answer correctly rounded-off is 25.925 (3D)

1.19 CHAPTER SUMMARY

In the chapter we have demonstrated the use of the hand-calculating machine in the basic arithmetical operations.

Because, in practice, 'real' problems deal with calculations involving inexact quantities, referred to in the text as 'rounded-off' numbers, we have shown the effect of such errors on the final results obtained from these calculations. Thus one is able to check the answer to see if it is to a satisfactory degree of accuracy.

From § 1.11, for Addition and Subtraction, there is the resultant absolute error modulus given by

$$|a_3| \leqslant |a_1| + |a_2| \tag{2}$$

From § 1.14 and § 1.16, in Multiplication and Division, the same formula is shown to satisfy both, giving the absolute relative error modulus as

$$\left|\frac{a_3}{(A_r)_3}\right| \leqslant \left|\frac{a_1}{(A_r)_1}\right| + \left|\frac{a_2}{(A_r)_2}\right| \tag{3}$$

Following the explanation of how to extract Square Roots in § 1.17, § 1.18 considers Powers and Roots, where the effect of the 'rounding-off' is shown to satisfy the formula

$$\left|\frac{a_2}{(A_r)_2}\right| = |p| \cdot \left|\frac{a_1}{(A_r)_1}\right| \tag{4}$$

where 'p' is the power concerned in the calculation $\{(A_r)_1\}^p$

1.20 EXAMPLES

1. Solve for x the equation $0{\cdot}138x = \dfrac{6{\cdot}96 \times 1{\cdot}063}{6{\cdot}78}$

2. Evaluate $\sqrt{\{(9{\cdot}485)^2 - (3{\cdot}615)^2\}}$

3. Evaluate $\dfrac{2{\cdot}69}{0{\cdot}386} + \dfrac{7{\cdot}536}{6{\cdot}87}$

For practice and experience, use both 'build-up' division and 'tear-down' division for this question. One method will act as a check on the other. Work to 4 dec. places.

4. In the following calculations all the quantities are subject to rounding-off errors.

a $1{\cdot}386 - 0{\cdot}987 + 7{\cdot}6485$
b $1{\cdot}7418 \times 8{\cdot}37$
c $15{\cdot}74 \div 3{\cdot}6976$

Calculate the absolute error modulus in each case and hence give each answer to a satisfactory degree of accuracy.

5. Calculate the maximum transmitted error to the value of x in Question 1 above, if all the numerical values are rounded-off.

6. Evaluate $\sqrt{\left\{\dfrac{6{\cdot}2343 \times 0{\cdot}82137}{2{\cdot}7268}\right\}}$ and the maximum transmitted error if the given numbers are rounded-off to five significant figures.

(A.E.B.)

7. If all the factors of $\dfrac{16{\cdot}35}{0{\cdot}864 \times 30{\cdot}8725}$ are rounded-off give the answer as accurately as is reasonable. Give also the absolute error modulus for this evaluation.

8. Evaluate

$$\frac{17{\cdot}678}{3{\cdot}471} + (9{\cdot}617)^2 \div (3{\cdot}716 \times 1{\cdot}85)$$

9. Define the terms absolute error; relative error and absolute error modulus. Obtain the range of values within which the exact value of

$$2{\cdot}7654 + 3{\cdot}8006 - \frac{15{\cdot}178}{0{\cdot}9876}$$

lies, if all the numerical quantities are 'rounded-off'.

10. A cylindrical bolt has a radius of 0·274 cm (3D) and length 3·45 cm (2D). Taking $\pi = 3{\cdot}142$ (3D) estimate the absolute error in calculating the volume of the bolt from this information and hence give the volume to an acceptable degree of accuracy.

11. (a) Loss of significant figures occurs when nearly equal numbers are subtracted. Illustrate this by considering the solutions of

$$x^2 - 2ax + b = 0 \text{ where } a^2 > b$$

Find as accurately as possible the roots of the above equation when

$$a = 2 \text{ exactly and}$$
$$b = 0{\cdot}0001 \pm 0{\cdot}000\,05$$

(b) Find the absolute error in

$$\frac{a}{b} + \frac{bc}{d} \text{ when } a = d = 0{\cdot}75 \pm 0{\cdot}006$$

$$\text{and} \quad b = \frac{c}{2} = 12{\cdot}35 \pm 0{\cdot}005$$

(Cambridge).

12. What is meant by absolute and relative errors?

If $z = (0{\cdot}51x + 1{\cdot}73)/(x + 0{\cdot}45)$

and the 2D coefficients are in error by $\pm\,0{\cdot}005$, find z together with its absolute and relative errors when $x = 0{\cdot}4 \pm 0{\cdot}1$

(Cambridge).

2

Programming Calculations

2.1 THE NEED FOR FLOW CHARTS

Although most calculations can be carried out in different ways, one way may be easier and quicker to use, or give greater accuracy, than others. Bearing in mind the desired accuracy it is important that some thought is first given as to the easiest way to carry out a set of calculations which are to be used repeatedly. The easiest way is always the best as mistakes are then less likely to be made.

The term 'programming' is used here to describe the process of arranging into a logical sequence the calculations to be carried out. This sequence may be described by the use of instructions, a list of such instructions forming a 'programme'. The programme may be left in this form for application to hand calculating machines, but if it is to be used for an electronic computer then it will be necessary to translate the instructions into a form which it can accept. However, for both types of machine the initial list of instructions is made clearer by the use of a flow chart.

2.1.1 Flow charts

A flow chart can be used to clarify any list of instructions which must be carried out in order. Basically the flow chart consists of writing each instruction in a 'box'. For example 'Unlock the door and open it'.

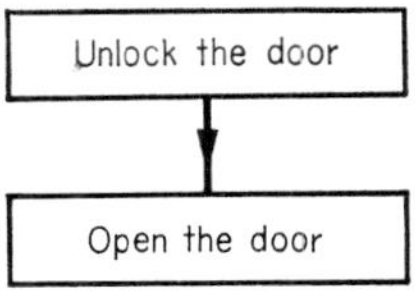

The boxes are connected to each other by arrows showing the order in which the instructions have to be obeyed. This order is usually important, as in the above example and the following recipe for making pastry:

'Sieve 8 oz. flour and a pinch of salt into a mixing bowl. Rub in 4 oz. fat very lightly and mix with a knife with just enough cold water to make the pastry bind together'.

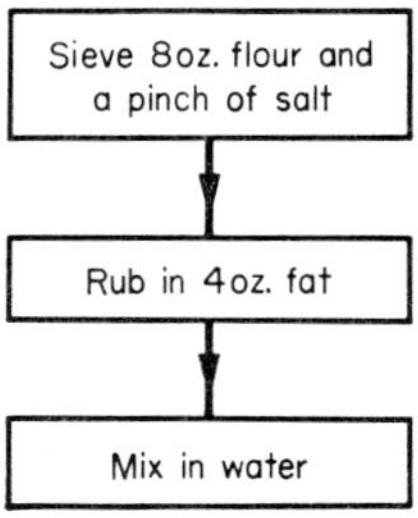

The instructions in a flow chart may be of several different types, and for each type a different symbol is used. For example an instruction may be in the form of a question having the two possible answers yes or no. In this case we use the symbol:

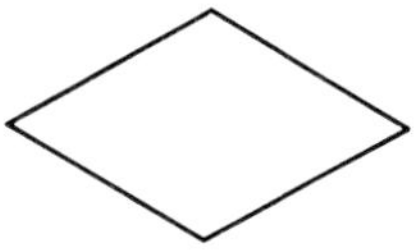

which is called the 'decision' symbol.

For example

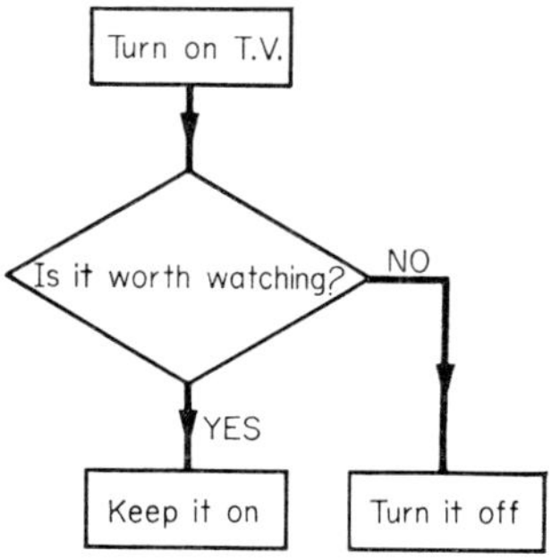

In this example the decision symbol is used when the instruction separates the two possible conclusions which can be reached. In the type of calculations covered in this book the decision symbol is usually used to form a loop of instructions which may be obeyed several times.

For example, in the following flow chart the first three instructions form a loop.

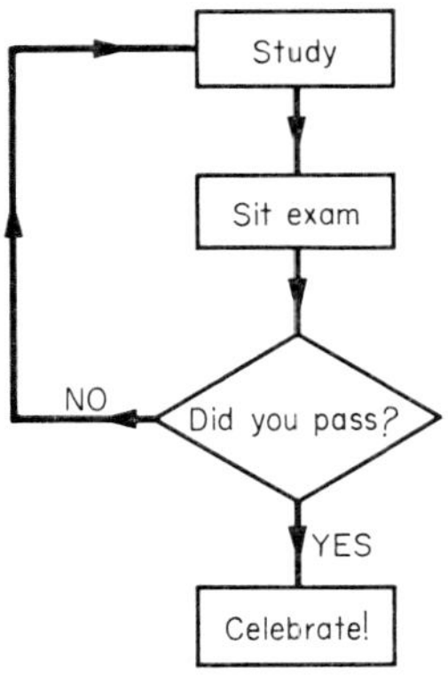

The following symbol may be used to indicate the first and last instructions in the flow chart and is called the 'terminal' symbol:

Using this symbol the above chart is then as follows:

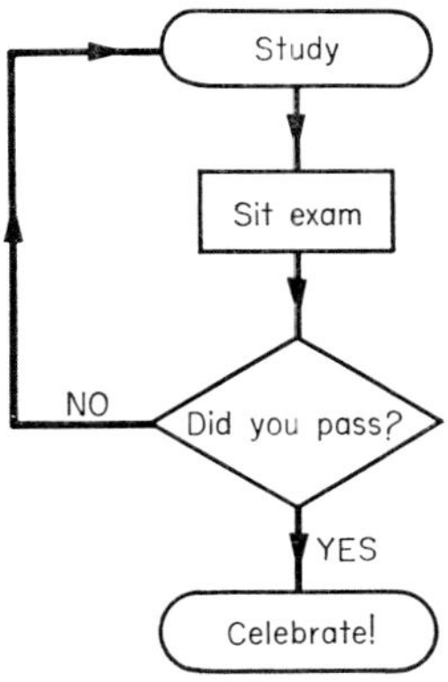

2.1.2

The following are some typical examples of flow charts.

Worked example 1

To find the *n*th. power of a number using a calculating machine.

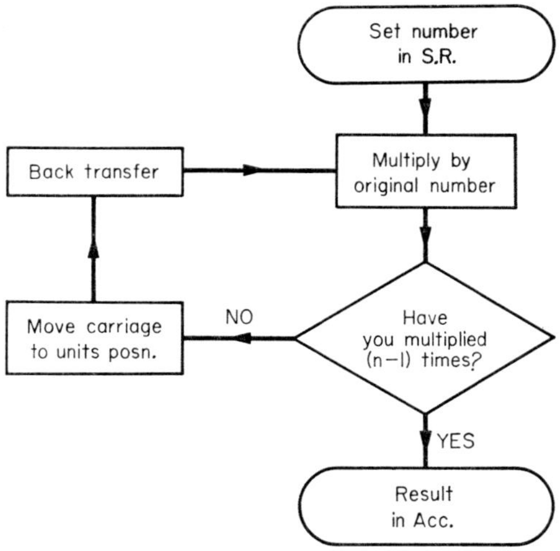

Note

If there are N digits in the number, the S.R. capacity must be at least $(n - 1) \times N$ places, and if there are N′ digits after the decimal point the number must be set with $(n - 1) \times N'$ decimal places in the S.R., otherwise the capacity of the S.R. will be exceeded and the programme would have to be modified to take account of this. For example, in calculating 41^3 the S.R. must have a capacity of four figures as 41^2 is 1681. In calculating $(8{\cdot}2)^4$ the S.R. must have a capacity of six figures and three decimal places should be set as $(8{\cdot}2)^3$ is 551·368. These rules give the maximum number of figures required and in many cases the number used will be less.

Worked example 2

Division by method B (1.15.2).

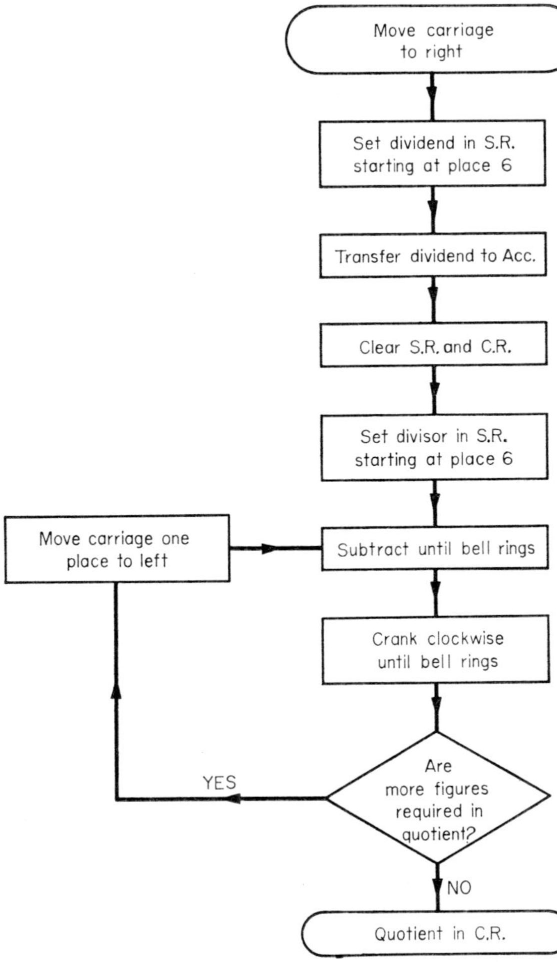

Worked example 3

To find the roots of the quadratic equation $ax^2 + bx + c = 0$, $a \neq 0$.

This chart differs from the last two in that it shows only the general method of solution and does not break it down into steps which can be performed in one operation of a hand calculating machine. This form of flow chart is the type used in the first stage of writing a programme for a digital computer.

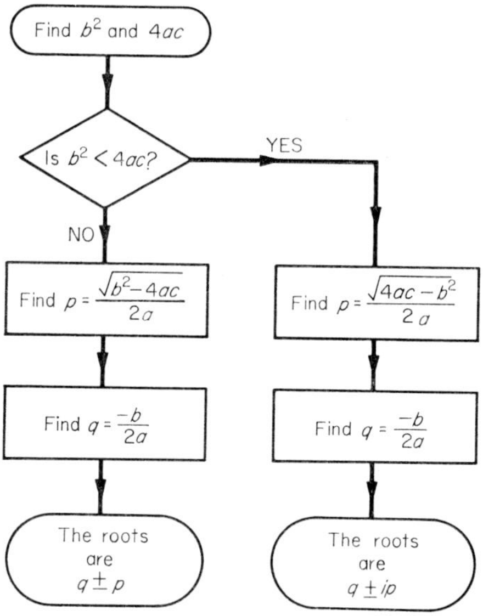

Since the instruction 'Find $q = -b/2a$' occurs in both the branches of this chart it could be placed before the decision step, thus making the chart more concise. See other examples in § 1.6.2 and § 1.17.3.

2.1.3 Examples

Draw flow charts to show methods of evaluating the following, using a hand calculating machine:

1. $11{\cdot}452 - 8{\cdot}614 - 5{\cdot}127$
2. $(7{\cdot}28 + 3{\cdot}641) \times 0{\cdot}92$
3. Division by method C (1.15.3), build-up division.
4. Volume of a cone, $\frac{1}{3}\pi r^2 h$.
5. Third side of a triangle by the cosine rule.
6. Complete the flow chart for the solution of a quadratic equation.

2.2 SOME STANDARD NOTATION

2.2.1 The function notation

To save writing a long expression repeatedly the shorthand notation $f(x)$ can be used to denote a function of x. Similarly $f(y)$ represents a function of y.

The value of the function at a particular value of x, $x = a$ say, is denoted by $f(a)$.

For example if $f(x) \equiv x^2 + 2x + 6$

then $f(0) = 6$ and $f(-3) = 9$

The equation $x^2 + 2x + 6 = 0$ can be represented by $f(x) = 0$.

As well as representing a particular function at one time, $f(x)$ can be used to represent any function of x in general. This is particularly useful in much theoretical work and in writing down formulae concisely (e.g. That for Newton's method in § 4.6).

If more than one function is being discussed at a time then the different functions can be described by using suffices or other letters. For example $f_1(x), f_2(x), F(x)$ and $g(x)$ could represent four different functions of x.

The first derivative of a function is denoted by $f'(x)$, the second by $f''(x)$ etc., using small roman numerals for fourth and higher derivatives. Thus, if $f(x) \equiv 6x^3 - 4x^2 + 2x + 8$ then $f'(x) \equiv 18x^2 - 8x + 2$ and $f''(x) \equiv 36x - 8$. Derivatives for particular values of x are written in the same way as for the original functions. For example here $f'(0) = 2$ and $f''(2) = 64$.

2.2.2 Values of the variable used in a table

A table of values of a function is usually given over a restricted range of values of the variable, at equal intervals of the variable. Thus a function of x may have to be tabulated from $x = 0{\cdot}5$ to $x = 1{\cdot}0$ for x increasing in steps of 0·1, i.e. for $x = 0{\cdot}5, 0{\cdot}6, 0{\cdot}7, 0{\cdot}8, 0{\cdot}9$ and 1·0. This can be put briefly as: 'Tabulate $f(x)$ for $x = 0{\cdot}5(0{\cdot}1)\ 1.0$.'

Tabulate $f(x)$ for $x = 0{\cdot}200(0{\cdot}001)0{\cdot}250$ would thus mean evaluating the function from $x = 0{\cdot}200$ to $x = 0{\cdot}250$, increasing x in steps of 0·001.

2.2.3 Examples

1. If $f(x) \equiv 6x^2 - 7x + 2$ write down (i) $f(0)$, (ii) $f(3)$, (iii) $f(-4)$, (iv) $f'(0)$ (v) $f'(3)$, (vi) $f''(x)$.
2. If $f(x) \equiv \cos x + \sin x$ write down (i) $f(0)$, (ii) $f(\pi)$, (iii) $f(-\pi/2)$, (iv) $f'(0)$. (v) What is the connection between $f(x)$ and $f''(x)$?

2.3 EVALUATION OF POLYNOMIALS BY NESTED MULTIPLICATION

In the course of obtaining solutions to some problems a polynomial must be evaluated. If the variable and coefficients are small integers the value of the polynomial can easily be found by evaluating each term and finding the sum of these results.

e.g. $3x^3 - 2x^2 + 3$ at $x = 2$ is $24 - 8 + 3 = 19$.

Often either the variable, or the coefficients, or both will not be small integers and this method is then too time consuming. The method usually used in such cases is that of 'nested multiplication' or 'nesting' which follows. Other methods are given in § 7.1.1 using synthetic division in which the same series of values is obtained but they are written down, and § 5.2 using differences to tabulate a polynomial.

Consider the value of the polynomial:

$$a_n x^n + a_{n-1}x^{n-1} + a_{n-2}x^{n-2} + \ldots\ldots + a_0 \text{ at } x = x_0$$

Take the first coefficient a_n and multiply by x_0. Add the second coefficient a_{n-1} to the result. Multiply by x_0 and add the third coefficient a_{n-2}. Continue this process until the last coefficient has been added *when the required value has been obtained.* The successive values that have been found are as follows:

$$a_n x_0 + a_{n-1}$$

$$(a_n x_0 + a_{n-1})x_0 + a_{n-2}$$

$$[(a_n x_0 + a_{n-1})x_0 + a_{n-2}]x_0 + a_{n-3}$$

" " " " " "

$$\{\ldots\ldots\ldots\ldots + a_1\}x_0 + a_0$$

e.g. The steps necessary to evaluate $4x^3 + 3x^2 - 2x + 1$ are:

1. $4x + 3$
2. $(4x + 3)x - 2$
3. $[(4x + 3)x - 2]x + 1$

The third step clearly gives the original polynomial.

Notes

1. Since rounding off errors will accumulate from each term of the sequence it is desirable to carry in the working at least two more places than the number required in the result.
2. Any missing powers of x must be included by setting the coefficient to zero (Otherwise for example $3x^3 + 2x^2 + 1$ would be evaluated as $3x^2 + 2x + 1$).
3. Working as shown above should not normally be written down, only the final result being noted.

In practice there are three different cases of the method, examples of which follow.

2.3.1 Case 1: x_0 and all the terms of the sequence are positive.

Worked example

(a) $1{\cdot}84x^3 + 2{\cdot}05x^2 + 1{\cdot}42x + 3{\cdot}89$ at $x = 2{\cdot}34$ to 3D.

Here the coefficients are 1·84, 2·05, 1·42 and 3·89 so that the nested multiplication sequence is,

1. $1{\cdot}84 \times 2{\cdot}34 + 2{\cdot}05$
2. $(1{\cdot}84 \times 2{\cdot}34 + 2{\cdot}05) \times 2{\cdot}34 + 1{\cdot}42$
3. $[(1{\cdot}84 \times 2{\cdot}34 + 2{\cdot}05) \times 2{\cdot}34 + 1{\cdot}42] \times 2{\cdot}34 + 3{\cdot}89$.

Set decimal markers as follows: S.R. 5D C.R. 2D Acc. 7D.

The following table shows the contents of each register after each operation has been carried out.

	S.R.	C.R.	Acc.
Set 1·84 in S.R. and multiply by 2·34	1·840 00	2·34	4·305 600 0
Return carriage to units position (i.e. here place 3) Set 2·05 in S.R. and add	2·050 00	3·34	6·355 600 0
Back transfer	6·355 60	0	0
Multiply by 2·34	6·355 60	2·34	14·872 104 0
Return carriage to units position Set 1·42 in S.R. and add	1·420 00	3·34	16·292 104 0
Back transfer	16·292 10	0	0
Multiply by 2.34	16·292 10	2·34	38·123 514 0
Return carriage to units position Set 3·89 in S.R. and add	3·890 00	3·34	42·013 514 0

i.e. required value is 42·014.

Note the basic sequence of operations which is repeated:

Multiply, Move carriage, Add, Back transfer.

This is followed in each of the three cases. These operations link together as follows:

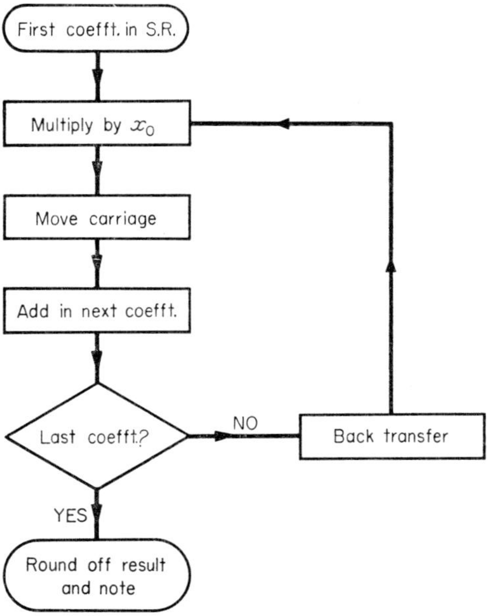

(b) One coefficient zero.

Worked example

(b) $1{\cdot}74x^3 + 3{\cdot}12x + 0{\cdot}86$ at $x = 0{\cdot}6441$ to 3D. S.R. 5D C.R. 4D Acc. 9D.

	S.R.	C.R.	Acc.
Set 1·74 in S.R. and multiply by ·6441	1·740 00	0·6441	1·120 734 000
Return carriage to units position (place 5)			
Back transfer (i.e. zero added)	1·120 73	0	0
Multiply by ·6441	1·120 73	0·6441	0·721 862 193
Return carriage to units position			
Set 3·12 in S.R. and add	3·120 00	1·6441	3·841 862 193
Back transfer	3·841 86	0	0
Multiply by ·6441	3·841 86	0·6441	2·474 542 026
Return carriage to units position			
Set 0·86 in S.R. and add	0·860 00	1·6441	3·334 542 026

i.e. required value is 3·334.

2.3.2 Case 2 x_0 positive but one or more terms of the sequence negative.

Worked example

(c) $1{\cdot}03x^3 - 2{\cdot}44x^2 - 1{\cdot}63x + 3{\cdot}89$ at $x = 2{\cdot}345$ to 3D.
S.R. 5D C.R. 3D Acc. 8D

	S.R.	C.R.	Acc.
Set 1·03 in S.R. and multiply by 2·345	1·030 00	2·345	2·415 350 00
Return carriage to units position (place 4) Set 2·44 in S.R. and subtract	2·440 00	1·345	9..9·975 350 00
Back transfer and subtract to give magnitude	9..9·975 35	1·000	0·024 650 00
Back transfer	0·024 65	0	0
Multiply by 2·345	0·024 65	2·345	0·057 804 25
Return carriage to units position. Back transfer and subtract	0·057 80	1·000	9..9·942 200 00
Set 1·63 in S.R. and subtract	1·630 00	2·000	9..98·312 200 00
Back transfer and subtract	9..98·312 20	1·000	1·687 800 00
Back transfer	1·687 80	0	0
Multiply by 2·345	1·687 80	2·345	3·957 891 00
Return carriage to units position. Back transfer and subtract	3·957 89	1·000	9..96·042 110 00
Set 3·89 in S.R. and add	3·890 00	0	9..9·932 110 00
Back transfer and subtract to give magnitude	9..9·932 11	1·000	0·067 890 00

i.e. required value is — 0·068.

Note

Each time a negative value occurs after adding or subtracting a coefficient it is back transferred, subtracted, and back transferred to set its magnitude in the S.R. It is then multiplied by the variable and the correct sign of the product obtained by back transferring and subtracting.

Alternative methods

1. After obtaining the magnitude of a negative result treat this as being positive and reverse the signs of successive coefficients, e.g. in the above example after line 6 return carriage to units position and *add* 1·63. It is often difficult to remember what sign changes there have been with this method.
2. The obvious method is to use the negative value to form the product but there are cases when there will be insufficient guarding 9's and the final value shown will be incorrect. For this reason the method described is preferable at all times.

2.3.3 Case 3 x_0 negative

Worked example

(d) $1{\cdot}46x^3 + 2{\cdot}34x^2 + 2{\cdot}13x + 1{\cdot}29$ at $x = -1{\cdot}432$ to 3D.

This is the same problem as that of evaluating $-1{\cdot}46x^3 + 2{\cdot}34x^2 - 2{\cdot}13x + 1{\cdot}29$ at $x = 1{\cdot}432$. i.e. Change the sign of the coefficients of all odd powers of x in the polynomial, take the modulus of x_0 and then proceed as in previous examples.

S.R. 5D C.R. 3D Acc. 8D

	S.R.	C.R.	Acc.
Set 1·46 in S.R.	1·460 00	0	0
Negative multiply by 1·432	1·460 00	1·432	9..97·909 280 00
Return carriage to units position. Set 2·34 in S.R. and add	2·340 00	0·432	0·249 280 00
Back transfer	0·249 28	0	0
Multiply by 1·432	0·249 28	1·432	0·356 968 96
Return carriage to units position. Set 2·13 in S.R. and subtract	2·130 00	0·432	9..98·226 968 96
Back transfer and subtract	9..98·226 96	1·000	1·773 040 00
Back transfer	1·773 04	0	0
Multiply by 1·432	1·773 04	1·432	2·538 993 28
Return carriage to units position. Back transfer and subtract	2·538 99	1·000	9..97·461 010 00
Set 1·29 in S.R. and add	1·29	2·000	9..98·751 010 00
Back transfer and subtract	9..98·751 01	1·000	1·248 990 00

Thus the required value is $-1{\cdot}249$.

2.3.4 Accuracy of the evaluation of a polynomial

If either the coefficients, or the variable, or both are rounded off then there will be an error in the value obtained for the polynomial. This error can be found by calculating the errors in each term using the results obtained in the previous chapter.

Worked example

Evaluate $3{\cdot}128x^3 + 1{\cdot}672x^2 - 0{\cdot}989x + 4{\cdot}215$ at $x = 2{\cdot}23$ where both the coefficients and the variable are rounded off.

The relative error in $x^3 \leqslant 3 \times 0{\cdot}005/2{\cdot}23$ (By chapter 1 equation (4) see § 1.18)

$$< 0{\cdot}008$$

thus using chapter 1 equation (3) see § 1.14, the relative error in

$$3{\cdot}128x^3 < 0{\cdot}008 + \frac{0{\cdot}0005}{3{\cdot}128}$$

$$< 0{\cdot}009$$

so that the absolute error in $3{\cdot}128x^3 < 0{\cdot}009 \times 35$

$$\simeq 0{\cdot}3$$

Similarly the relative error in $x^2 \leqslant 2 \times \dfrac{0{\cdot}005}{2{\cdot}23}$

$$< 0{\cdot}005$$

and thus the relative error in $1{\cdot}672x^2 < 0{\cdot}005 + \dfrac{0{\cdot}0005}{1{\cdot}672}$

$$< 0{\cdot}006$$

so that the absolute error in $1{\cdot}672x^2 < 0{\cdot}006 \times 8$

$$< 0{\cdot}05$$

The relative error in $0{\cdot}989x < \dfrac{0{\cdot}0005}{0{\cdot}989} + \dfrac{0{\cdot}005}{2{\cdot}23}$

$$< 0{\cdot}003$$

so that the absolute error in $0{\cdot}989x < 0{\cdot}003 \times 2{\cdot}2$

$$< 0{\cdot}007$$

The absolute error in $4{\cdot}215 \leqslant 0{\cdot}0005$

Thus the absolute error in the evaluation of this polynomial at this value

$< 0{\cdot}3 + 0{\cdot}05 + 0{\cdot}007 + 0{\cdot}0005$

$< 0{\cdot}4$.

Now evaluating the polynomial to 3D we have $f(2{\cdot}23) = 45{\cdot}047$ so that $44{\cdot}6 < f(2{\cdot}23) < 45{\cdot}4$ and the value of the polynomial can be correctly rounded off to 45.

Note that here the first term contributes the main part of the error but that this is not necessarily true if the value of the variable is exact.

This example shows the errors that occur in the worst possible case when both coefficients and variable have been rounded off and maximum errors are considered at each stage, so that the true error is probably smaller than the estimate obtained here.

2.3.5 Examples

1. Evaluate the following polynomials to 3D for the given values of x.

(i)	$2{\cdot}3x^3 - 0{\cdot}6x^2 + 1{\cdot}8x - 2{\cdot}2$	$1{\cdot}6$
(ii)	$1{\cdot}64x^3 + 0{\cdot}37x - 0{\cdot}51$	$1{\cdot}61$
(iii)	$2{\cdot}713x^3 - 1{\cdot}325x^2 + 1{\cdot}764$	$0{\cdot}856$
(iv)	$1{\cdot}12x^3 - 0{\cdot}98x^2 + 1{\cdot}32x - 0{\cdot}74$	$0{\cdot}42$
(v)	$0{\cdot}78x^3 - 1{\cdot}03x - 1{\cdot}46$	$0{\cdot}88$
(vi)	$0{\cdot}2488x^3 - 0{\cdot}6814x^2 + 0{\cdot}5013$	$-0{\cdot}758$
(vii)	$8{\cdot}806x^4 - 7{\cdot}439x + 2{\cdot}499x - 8{\cdot}133$	$-0{\cdot}744$

2. Evaluate $3{\cdot}76x^3 - 5{\cdot}24x^2 + 1{\cdot}38x + 2{\cdot}32$ as accurately as possible at $x = 0{\cdot}487$ if the coefficients are rounded off but the value of x is exact.

3. Evaluate $2{\cdot}82x^3 - 1{\cdot}97x^2 - 3{\cdot}11x + 0{\cdot}25$ as accurately as possible at $x = 0{\cdot}513$ if the coefficients are exact but the value of x is rounded off.

4. Evaluate $6{\cdot}123x^3 + 2{\cdot}987x^2 - 5{\cdot}755x + 7{\cdot}204$ as accurately as possible at $x = -2{\cdot}603$ if both the coefficients and the variable are rounded off.

2.4 EVALUATION AND TABULATION OF OTHER FUNCTIONS

If any function is to be tabulated for several values of the variable some thought should first be given about the method to be used. This method must both ensure the desired accuracy and be applicable with the calculating aids available at the time. Provided that these conditions can be satisfied then ways should be sought of rearranging the computation in order to reduce the number of operations necessary, or so that the operations may be carried out as one continuous process on a calculating machine without intermediate writing down.

2.4.1 Rearranging computation

1. *To obtain the desired accuracy*

Some functions for certain values of the variable may be evaluated to different accuracies by different methods, so that a rearrangement of the chosen method may be necessary in order to obtain the desired accuracy.

Example (a). Some of the trigonometrical functions have rapid changes in value, or very small values, near to multiples of 90° or 180°. Thus if it is necessary to evaluate $1/\cos x$ it is more accurate to use tables of $\sec x$ if available, especially near to multiples of 90°. For example from four figure tables $\cos 1{\cdot}569 = 0{\cdot}001\,80$, which is only correct to 3S, and hence $1/0{\cdot}001\,80 = 555{\cdot}56$ is not $1/\cos 1{\cdot}569$ correct to 5S, whereas from tables $\sec 1{\cdot}569 = 556{\cdot}69$.

Example (b). In certain quadratic equations the usual method of solution results in a loss of accuracy in one of the roots. Other methods must therefore be devised for these cases and two such methods are given in §7.3.

2. *To simplify the working*

In the evaluation of the roots of an equation the working may be simplified by the choice of a suitable method. For example, to find the cube root of a number it is quicker to use the method given in §4.3 rather than devise a process using the method of §4.4.

In the evaluation of a function the working should be simplified, if possible, by rearranging the function itself. Functions involving trigonometrical, hyperbolic, exponential and logarithmic functions should be rearranged to the simplest form possible using such properties as $\cos^2 x + \sin^2 x = 1$, $2 \sinh x \cosh x = \sinh 2x$, $\log x^2 = 2 \log x$, $(e^x)^2 = e^{2x}$.

For example, $\sin^2 x \sec x$ can be rearranged as $\tan x \sin x$. The operations necessary for evaluating $\sin^2 x \sec x$ on a calculating machine are : (1) set $\sec x$, (ii) Multiply by $\sin x$, (iii) Move carriage to units position, (iv) Back transfer, (v) Multiply by $\sin x$. The corresponding operations necessary for evaluating $\tan x \sin x$ are: (i) Set $\sin x$, (ii) Multiply by $\tan x$. Both calculations involve looking up values in two tables. Thus there will be a large saving in time using this rearrangement if the function is to be tabulated.

Of the four basic operations division takes the longest time to carry out, so that the evaluation of a function will be speeded up if division can be replaced by multiplication. For example $(x^3 + \tan x)/\sin x$ can be rearranged as $\operatorname{cosec} x \,.\, (x^3 + \tan x)$ or $x^3 \operatorname{cosec} x + \sec x$ and $(1 + x^2)/e^x$ as $e^{-x}(1 + x^2)$ provided that the appropriate tables are available.

2.4.2 Functions involving powers or roots

If the function to be tabulated is of the form $[f(x)]^n$ the tabulation is best made in two or more steps, especially if n is rational. First make out a table of values of $f(x)$ (after simplifying $f(x)$ if possible) and then secondly tabulate $[f(x)]^n$. Due to the repetitive nature of the calculations for each step this method is usually quicker than evaluating $[f(x)]^n$ immediately after $f(x)$ for each value of x.

For equally spaced values of the variable it may be useful to calculate some differences, which can be used to provide a good approximation to each successive value of $[f(x)]^n$ by a similar method to that given in §5.2 for the tabulation of a polynomial.

2.4.3 Checking the tabulation of a function

The tabulation of a function may be checked either by differencing or by retabulation using another method which provides the same accuracy.

*Differencing does not indicate the precise correction which is needed in any incorrect value but will only show if and where any comparatively large mistake is made by disturbing the uniformity of the difference table. However, it can be a fairly quick method of checking. (See §5.1. and §5.3.3).

Retabulation by evaluating the function for all values of the variable by a different method provides an accurate check, but takes as long as the original tabulation.

Worked example

Evaluate $\dfrac{x^3 + \tan x}{\sin x}$ at $x = 1{\cdot}263$ to 4D.

This function may be rearranged to give either $\operatorname{cosec} x \,.\, (x^3 + \tan x)$ or $x^3 \operatorname{cosec} x + \sec x$. Both forms can be evaluated by a continuous process on a calculating machine as follows, using one guarding figure.
(i) $\operatorname{cosec} x(x^3 + \tan x)$. Calculate x^3, add $\tan x$ then multiply by $\operatorname{cosec} x$. S.R. 5D C.R. 5D Acc. 10D.

	S.R.	C.R.	Acc.
Set 1·263 in S.R. and multiply by 1·263	1·263 00	1·263 00	1·595 169 000 0
Return carriage to units position (place 6), back transfer and multiply by 1·263	1·595 16	1·263 00	2·014 687 080 0
Return carriage to units position clear C.R. and S.R.	0	0	2·014 687 080 0
Set $\tan x = 3{\cdot}145\ 65$ in S.R. and add	3·145 65	1	5·160 337 080 0
Back transfer	5·160 33	0	0
Multiply by $\operatorname{cosec} x = 1{\cdot}049\ 31$	5·160 33	1·049 31	5·414 785 872 3

Thus rounding off to 4D $\dfrac{x^3 + \tan x}{\sin x} = 5{\cdot}4148$

* These paragraphs may be omitted until the method of differencing is more fully studied in Chapter 5.

(ii) x^3 cosec x + sec x. Calculate x^3, multiply by cosec x then add sec x. S.R. 5D C.R. 5D Acc. 10D.

	S.R.	C.R.	Acc.
Set 1·263 in S.R. and multiply by 1·263	1·263 00	1·263 00	1·595 169 000 0
Return carriage to units position (place 6) back transfer and multiply by 1·263	1·595 16	1·263 00	2·014 687 080 0
Return carriage to units position, back transfer and multiply by cosec x = 1·049 31	2·014 68	1·049 31	2·114 023 870 8
Return carriage to units position, clear C.R. and S.R.	0	0	2·114 023 870 8
Set sec x = 3·300 77 in S.R. and add	3·300 77	1	5·414 793 870 8

Thus rounding off to 4D $\dfrac{x^3 + \tan x}{\sin x} = 5{\cdot}4148$, confirming the previous value.

*If tabulating this function use method (i) and, provided this tabulation is at equal intervals, it can be differenced and a few spot checks made. An alternative check would be to use method (ii). If in this case it was found that two values did not agree in the last decimal place, and that no mistakes had been made, then an extra guarding figure in the working would be necessary.

2.4.4 Accuracy of the evaluation of a function

Let $(A_r)_2 = f\{(A_r)_1\}$ be the approximation for the exact relationship $(A_e)_2 = f\{(A_e)_1\}$.

Then using $A_e = A_r - a$, and substituting in the exact relationship, we have $(A_r)_2 - a_2 = f\{(A_r)_1 - a_1\}$

giving $(A_r)_2 - a_2 = f\{(A_r)_1\} - a_1 f'\{(A_r)_1\}$ by Taylor's expansion†, neglecting terms involving powers of a_1.

But since $$(A_r)_2 = f\{(A_r)_1\}$$

it follows that $$a_2 = a_1 f'\{(A_r)_1\}$$

and thus $$|a_2| = |a_1| \,.\, |f'\{(A_r)_1\}| \qquad (5)$$

* May be omitted until method of differencing is more fully studied in Chapter 5.

† i.e. $f(a + h) = f(a) + hf'(a) + \dfrac{h^2}{2!}f''(a) + \ldots$

i.e. The absolute error modulus in the value of a function is equal to the product of the modulus of the first derivative of the function and the absolute error modulus of the variable.

The relative error modulus in the value of a function is obtained by dividing each side of (5) by $(A_r)_2 = f\{(A_r)_1\}$ respectively, giving

$$\left|\frac{a_2}{(A_r)_2}\right| = |\, a_1 \,| \, . \left|\frac{f'\{(A_r)_1\}}{f\{(A_r)_1\}}\right| \tag{6}$$

i.e. The relative error modulus in the value of a function is equal to the absolute error modulus of the variable multiplied by the modulus of the ratio of the first derivative to the function.

Note that if the function is a power or a root, i.e. $f\{(A_r)_1\} = \{(A_r)_1\}^p$ and thus $f'\{(A_r)_1\} = p\{(A_r)_1\}^{p-1}$, then substituting in (6) gives

$$\left|\frac{a_2}{(A_r)_2}\right| = |\, a_1 \,| \, . \left|\frac{p\{(A_r)_1\}^{p-1}}{\{(A_r)_1\}^p}\right|$$

so that

$$\left|\frac{a_2}{(A_r)_2}\right| = |\, p \,| \, . \left|\frac{a_1}{(A_r)_1}\right|$$

which is equation (4), (see § 1.18).

Worked Example 1

Evaluate sin(0·572) when 0·572 radian has been rounded off.
Here 0·572 has a maximum absolute error modulus of 0·0005. From six figure tables

$$f(0{\cdot}572) = \sin(0{\cdot}572) = 0{\cdot}541\,315$$

and

$$f'(0{\cdot}572) = \cos(0{\cdot}572) = 0{\cdot}840\,820$$

Thus using (5) the absolute error modulus in

$$\sin(0{\cdot}572) \leqslant 0{\cdot}0005 \times 0{\cdot}84$$
$$< 0{\cdot}0005$$

and hence the value of sin(0·572) is between 0·5408 and 0·5418. i.e. It can only be given correctly rounded off to 2D as 0·54 or may be given as 0·541 with an error of up to 1 in the last decimal place.

Worked example 2a

Evaluate $(1 + x^2)e^{-x}$ at $x = 1{\cdot}75$ which is exact.

Since x is exact $1 + x^2$ will be obtained exactly. If e^{-x} is found from six figure tables it will have an absolute error modulus of $0{\cdot}5 \times 10^{-6}$.
For $x = 1{\cdot}75$, $e^{-x} = 0{\cdot}173\,774$ and thus $(1 + x^2)e^{-x} = 0{\cdot}705\,956\,875$.
The relative error modulus in $(1 + x^2)e^{-x}$, (the same as the relative error modulus in e^{-x} since $1 + x^2$ is exact), is equal to $(0{\cdot}5 \times 10^{-6})/0{\cdot}17$, which is $< 2{\cdot}94 \times 10^{-6}$.

Hence the absolute error modulus in $(1 + x^2)e^{-x}$ is

$$< 2{\cdot}94 \times 10^{-6} \times 0{\cdot}7$$

which is about

$$2 \times 10^{-6}$$

Thus the value of $(1 + x^2)e^{-x}$ lies between 0·705 955 and 0·705 959 so that it may be rounded off to 0·705 96, i.e. to 5D.

Worked example 2b

Evaluate $(1 + x^2)e^{-x}$ at $x = 1{\cdot}75$ which has been rounded off.

(i) Relative error modulus in $(1 + x^2)$.
Since 1 is exact this is less than the relative error modulus in x^2,

i.e. from (4), § 1.18, $2 \times \dfrac{0{\cdot}5 \times 10^{-6}}{1{\cdot}75}$ which is $< 0{\cdot}6 \times 10^{-6}$.

(ii) Relative error modulus in e^{-x}.

Since $f(1{\cdot}75) = e^{-1{\cdot}75} = 0{\cdot}173\,774$
and $f'(1{\cdot}75) = -e^{-1{\cdot}75} = -0{\cdot}173\,774$

the modulus of the ratio of the first derivative to the function is 1 and thus using (6), in § 2.4.4, the relative error modulus in

$$e^{-x} = 0{\cdot}5 \times 10^{-2} \times 1$$
$$= 0{\cdot}5 \times 10^{-2}$$

Hence using (3), in § 1.14, the relative error modulus in

$$(1 + x^2)e^{-x} < 0{\cdot}5 \times 10^{-2}$$

giving the absolute error modulus in

$$(1 + x^2)e^{-x} < 0{\cdot}5 \times 10^{-2} \times 0{\cdot}7$$
$$< 0{\cdot}4 \times 10^{-2}$$
$$= 4 \times 10^{-3}$$

Thus the value of $(1 + x^2)e^{-x}$ lies between 0·702 and 0·710 so that it may be rounded off to 0·7, i.e. to 1D.

These examples show that it is necessary to determine the accuracy to which a function may be evaluated. The use of six-figure tables rarely implies that the final value can be quoted correct to six figures.

2.4.5 Examples (Using six figure tables)

1. Round off each of the following correctly given that the variable in each case is rounded off. (i) cos(1·162) (ii) log(2·69) (iii) $e^{-1{\cdot}728}$ (iv) tan(0·956).

2. To what number of decimal places may the following be evaluated at the exact value $x = 0{\cdot}75$. (i) $x \cos^2 x$ (ii) $\cos x \sin x$ (iii) $4x^2 + \cosh x$ (iv) $\sqrt{(x + \sin^{-1} x)}$.
3. Tabulate $(1 + x^2)\sin x$ for $x = 1{\cdot}00(0{\cdot}05)1{\cdot}50$ to 5D.
4. Tabulate $(\cos^2 x - x^2)/(\cos^2 x)$ for $x = 0(0{\cdot}01)0{\cdot}05$ to 6D.
5. Tabulate $2x + \log x^4$ for $x = 0{\cdot}2(0{\cdot}2)1{\cdot}2$ to 5D.

2.5 SUMMATION OF SERIES

Series arise in the course of certain methods used in numerical analysis e.g. (a) a series solution of a differential equation, (b) in the use of interpolation formulae. They can also be used to derive values of many functions, such as those tabulated in Chambers' Shorter Six-Figure Tables.

They are of most use when only the first few terms of the series will give the accuracy required, i.e. when they converge rapidly. If convergence is slow then either some other method not using the series, or some transformation of the series must be found.

There is often some connection between the terms of a series which may be used to simplify the evaluation of each term.

e.g. The series for e^x is $1 + x + x^2/2! + x^3/3! + x^4/4! + \ldots$ Here the third term, $x^2/2! = x/2 \,.\, x$, x being the second term; the fourth term $x^3/3! = x/3 \,.\, x^2/2!$, $x^2/2!$ being the third term; the fifth term $x^4/4! = x/4 \,.\, x^3/3!$ etc. so that in general the $(n + 1)$th term $= x/n$ times the nth term. i.e. Each term can be obtained from the previous term on multiplying it by a suitable factor instead of evaluating it independently. The first few factors, $x/2$, $x/3$, $x/4$, etc., could be found mentally here.

To evaluate a series to a given accuracy it is desirable to carry two guarding figures in the calculation of each term, though if convergence is rapid one such guarding figure will probably be sufficient. The terms must be evaluated until it is certain that the next one will not affect the final result. This usually implies that at least the first two terms which are smaller than the permissible error must be included.

Worked example 1

Evaluate $e^{0.15}$ correct to 6D, using $e^x = 1 + x + x^2/2! + \ldots$

Term 1.	1		1
2.	x		0·15
3.	$\dfrac{x^2}{2!} = \dfrac{x}{2} \cdot x$	$0{\cdot}075 \times 0{\cdot}15$	0·011 25
4.	$\dfrac{x^3}{3!} = \dfrac{x}{3} \cdot \dfrac{x^2}{2!}$	$0{\cdot}05 \times 0{\cdot}011\,25$	0·000 562 5
5.	$\dfrac{x^4}{4!} = \dfrac{x}{4} \cdot \dfrac{x^3}{3!}$	$0{\cdot}0375 \times 0{\cdot}000\,562\,5$	0·000 021 09

6.	$\frac{x^5}{5!} = \frac{x}{5} \cdot \frac{x^4}{4!}$	$0{\cdot}03 \times 0{\cdot}000\,021\,09$	$0{\cdot}000\,000\,63$
7.	$\frac{x^6}{6!} = \frac{x}{6} \cdot \frac{x^5}{5!}$	$0{\cdot}025 \times 0{\cdot}000\,000\,63$	$0{\cdot}000\,000\,02$
8.	$\frac{x^7}{7!} = \frac{x}{7} \cdot \frac{x^6}{6!}$	$0{\cdot}021\,428\,57 \times 0{\cdot}000\,000\,02$	$0{\cdot}000\,000\,00$
			$\Sigma = 1{\cdot}161\,834\,24$

Thus to 6D $e^{0.15} = 1{\cdot}161\,834$.

In practice it is only necessary to write down the column giving the values of the terms. If the terms of the series alternate in sign then note the terms in two columns according to sign. The terms are evaluated up to the eighth which is the second one smaller than 0·000 000 5. It will be seen that in this example one guarding figure would have been sufficient.

Although the series for e^x converges for all values of x, (i.e. has a finite value for all finite values of x) convergence is slow when $|x| > 1$ and it is then easier to resolve the function into two or more factors, e.g. $e^{2{\cdot}5} = e \,.\, e \,.\, e^{0{\cdot}5}$.

Worked example 2

Evaluate $$\sum_{n=1}^{\infty} \frac{1}{3^n(n+1)} \text{ correct to 3D.}$$

$$\sum_{n=1}^{\infty} \frac{1}{3^n(n+1)} = \frac{1}{3.2} + \frac{1}{9.3} + \frac{1}{27.4} + \frac{1}{81.5} + \ldots$$

which can be rewritten as:

$$\frac{1}{3.2} + \frac{2}{3.3}\left(\frac{1}{3.2}\right) + \frac{3}{3.4}\left(\frac{1}{9.3}\right) + \frac{4}{3.5}\left(\frac{1}{27.4}\right) + \ldots$$

$$= \frac{1}{6} + \frac{2}{9}\left(\frac{1}{3.2}\right) + \frac{3}{12}\left(\frac{1}{9.3}\right) + \frac{4}{15}\left(\frac{1}{27.4}\right) + \ldots$$

in order to obtain each term from the previous term.

Term 1.	1/6	0·166 67
2.	(2 × 0·166 67) ÷ 9	0·037 04
3.	(3 × 0·037 04) ÷ 12	0·009 26
4.	(4 × 0·009 26) ÷ 15	0·002 47
5.	(5 × 0·002 47) ÷ 18	0·000 69
6.	(6 × 0·000 69) ÷ 21	0·000 20
7.	(7 × 0·000 20) ÷ 24	0·000 06
		Σ = 0·216 39

Thus to 3D $$\sum_{n=1}^{\infty} \frac{1}{3^n(n+1)} = 0{\cdot}216$$

In the above rearrangement of the series each term is written as the product of the previous term and a fraction. Each term is then calculated by multiplying the previous term by the numerator of this fraction and then dividing the result by the denominator. Tear down division is used as the dividend is then already set in the accumulator. Alternatively build-up division may be used if each term if firstly divided by the denominator of the fraction so that the quotient is obtained in the setting register and can then be multiplied directly by the numerator.

Note that the sum of the series is found up to the seventh term, the second term which is smaller than 0·0005.

2.5.1 Examples

1. Evaluate $\sum_{n=1}^{\infty} \frac{1}{3^n n}$ correct to 3D.

2. Evaluate sin(0·15) correct to 6D.

$$\sin x = x - \frac{x^3}{3!} + \frac{x^5}{5!} - \frac{x^7}{7!} + \ldots \text{ for all } x.$$

3. Evaluate cos (0·45) correct to 6D.

$$\cos x = 1 - \frac{x^2}{2!} + \frac{x^4}{4!} - \frac{x^6}{6!} + \ldots \text{ for all } x.$$

4. Evaluate sinh (0·16) correct to 6D.

$$\sinh x = x + \frac{x^3}{3!} + \frac{x^5}{5!} + \ldots \text{ for all } x.$$

5. Evaluate cosh (0·21) correct to 6D.

$$\cosh x = 1 + \frac{x^2}{2!} + \frac{x^4}{4!} + \ldots \text{ for all } x.$$

3

Sketches of Simple Functions

3.1 CURVE SKETCHING—SKETCHES OF SIMPLE FUNCTIONS

A good sketch of a curve is a neat and careful drawing showing the essential features and properties of the curve. To do this it is not necessary to produce an accurate graph, but it is often surprising to see how accurately and quickly the main parts of a curve can be drawn using the minimum amount of calculation.

Quick sketches are required in many types of mathematical problem, e.g. in locating roots of equations, determining limits of integration. Hence, the ability to produce a useful sketch quickly is well worth acquiring. The graphs of many of the simpler functions should be remembered and can become a pictorial (and very economical) method of recalling the properties of these functions. All the functions discussed in this chapter are required and will be encountered again and again in later parts of the book.

A knowledge of calculus is not essential* in this chapter, which is intended to be read before, and indeed forms a useful preliminary to, the introduction of calculus into a mathematics course.

However, curve sketching and calculus should not be kept apart indefinitely and on a second reading at a later stage the reader could simplify some parts of the work when he is able to use calculus as well as the methods outlined here.

It must be emphasised that throughout this chapter we are dealing only with real values of the variables, i.e. no point can be plotted on any of these graphs if its co-ordinates involve the square root of a negative number.

3.1.1 Simple polynomials

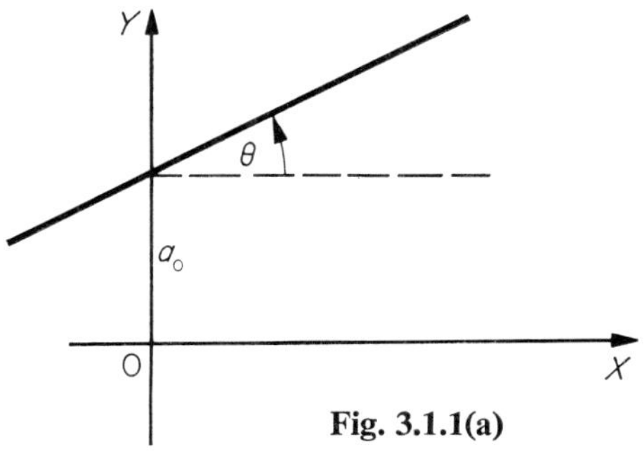

Fig. 3.1.1(a)

*Some use is made of calculus in §§ 3.1.1(d) and 3.5.

(a) *The straight line* $y = a_0 + a_1 x \quad a_1 \neq 0$, when $x = 0$, $y = a_0$.

The gradient $(\tan \theta) = a_1$ (See Fig. 3.1.1(a))

(b) *The parabola* $y = a_0 + a_1 x + a_2 x^2 \equiv a_2(x - \alpha)(x - \beta)$, shown in Fig. 3.1.1 (b), (where $x = \alpha$, and $x = \beta$ are the roots of the quadratic equation $a_2 x^2 + a_1 x + a_0 = 0$).

(i) if a_2 is +ve and $\alpha < \beta$ y is −ve for $\alpha < x < \beta$. $y \to +\infty$ when $x \to \pm\infty$

(ii) if a_2 is −ve and $\alpha < \beta$ y is +ve for $\alpha < x < \beta$. $y \to -\infty$ when $x \to \pm\infty$

(iii) There are other cases, namely those involving $\alpha = \beta$ or α and β complex. However, these are not illustrated here.

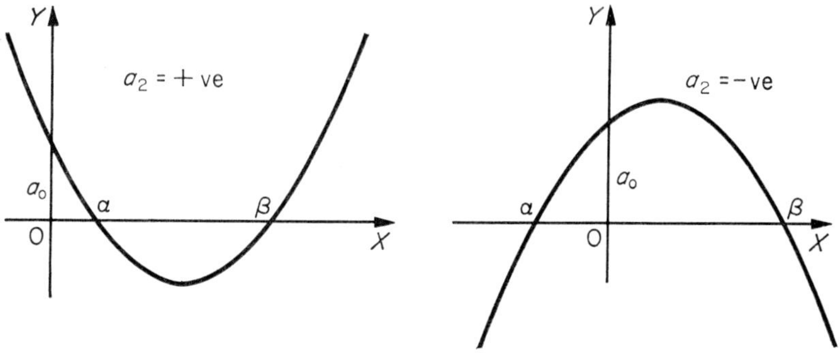

Fig. 3.1.1(b)

(c) *The cubic* $y = a_0 + a_1 x + a_2 x^2 + a_3 x^3 \equiv a_3(x - \alpha)(x - \beta)(x - \gamma)$, $a_3 \neq 0$, as shown in Fig. 3.1.1 (c) (where α, β, γ are the roots of the cubic equation $a_3 x^3 + a_2 x^2 + a_1 x + a_0 = 0$).

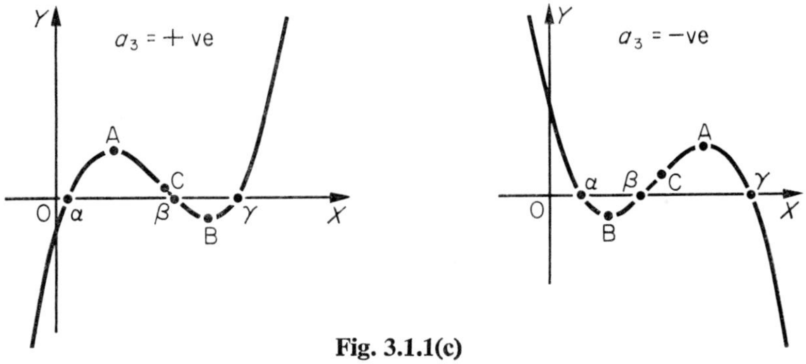

Fig. 3.1.1(c)

(i) if a_3 is +ve and $\alpha<\beta<\gamma$ y is +ve for $\alpha<x<\beta$ and $x<\gamma$

(ii) if a_3 is −ve $y=$ +ve for $x<\alpha$ and $\beta<x<\gamma$

There is a maximum (A) and a minimum (B) and midway between them there is a point of inflexion (C) i.e. a point where the tangent at the point crosses the curve, which is also the centre of symmetry of the curve.

(iii) There are other cases, namely those involving two of α, β, γ being equal or complex. However these are not illustrated here.

(d) *Algebraic polynomials* (May be omitted on first reading).
The polynomial of degree n has the equation

$$y = a_0 + a_1x + a_2x^2 + \ldots\ldots + a_nx^n,\ a_n \neq 0 \tag{1}$$

The graph of this function has the following properties:

1. No straight line can cross it at more than n points. This is true because the equation of a straight line can be written in the form

$$y = mx + c \tag{2}$$

and the solution of equations (1) and (2) gives the co-ordinates of the points of intersection. The number of these intersections is equal to the number of real roots of the equation

$$mx + c = a_0 + a_1x + \ldots\ldots + a_nx^n$$

which gives the x co-ordinates of the points of intersection. As this equation cannot have more than n roots, there cannot be more than n points of intersection.

2. The curve is continuous (i.e. consists of one unbroken line) from $x = -\infty$ to $x = +\infty$.

3. The total number of maximum and minimum points on the curve cannot be greater than $n-1$ and is frequently less than $n-1$. This follows because dy/dx is of degree $n-1$ and the equation $(dy/dx) = 0$ cannot have more than $n-1$ real roots. Furthermore some points at which $(dy/dx) = 0$ may prove to be points of inflexion.

4. Similarly there will not be more than $n-2$ points of inflexion.

5. (a) The curve passes through the point $(0, a_0)$ with gradient a_1, i.e. the line $y = a_0 + a_1x$ is a tangent to the curve at $x = 0$.

(b) The curve is above or below this tangent according to the sign of the next term in the polynomial. [§ 1.2.2]

6. When a_n is +ve, $y \to +\infty$ as $x \to +\infty$ (all n)

$y \to +\infty$ as $x \to -\infty$ if n is even

$y \to -\infty$ as $x \to -\infty$ if n is odd.

Worked example 1

$y = x^4 - 5x^2 + 4 \equiv (x+2)(x+1)(x-1)(x-2)$

(i) $y = 0$ when $x = -2, -1, +1, +2$ and $y = 4$ when $x = 0$

(ii) $y \to +\infty$ when $x \to \pm\infty$

(iii) y is +ve when $x > 2$ and when $-1 < x < +1$ and $x < -2$
y is −ve when $+1 < x < +2$ and when $-2 < x < -1$.

Hence the graph is of the form as shown in Fig. 3.1.1(d)

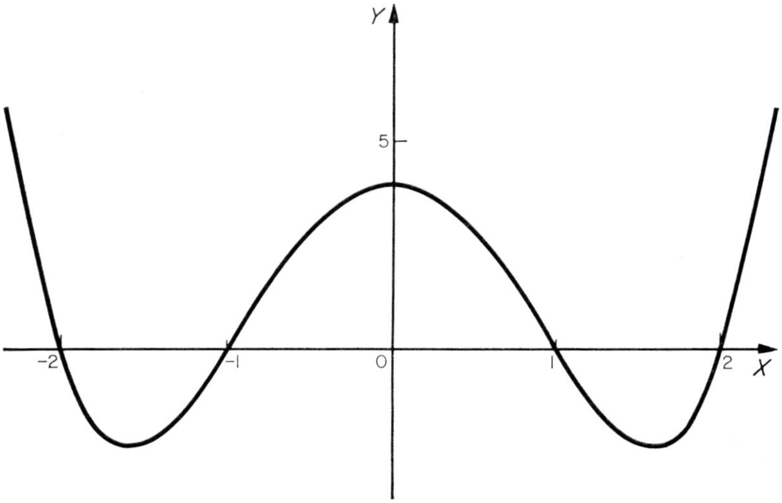

Fig. 3.1.1(d)

Worked example 2

$y = (x^2 - 3)(x - 1)^2 = +x^4 \ldots\ldots\ldots\ldots -3$

when $y = 0$, $x = 1, 1, +\sqrt{3}, -\sqrt{3}$, and when $x = 0$, $y = -3$

$(x - 1)^2$ is +ve for all x (except zero at $x = 1$)

$x^2 - 3$ is +ve (i.e. y is +ve) for $|x| > \sqrt{3}$

$x^2 - 3$ is −ve (i.e. y is −ve) for $|x| < \sqrt{3}$

The coefficient of x^4 is +ve $\therefore y \to +\infty$ as $x \to \pm\infty$

Therefore the graph is of the form as shown in Fig. 3.1.1(e).

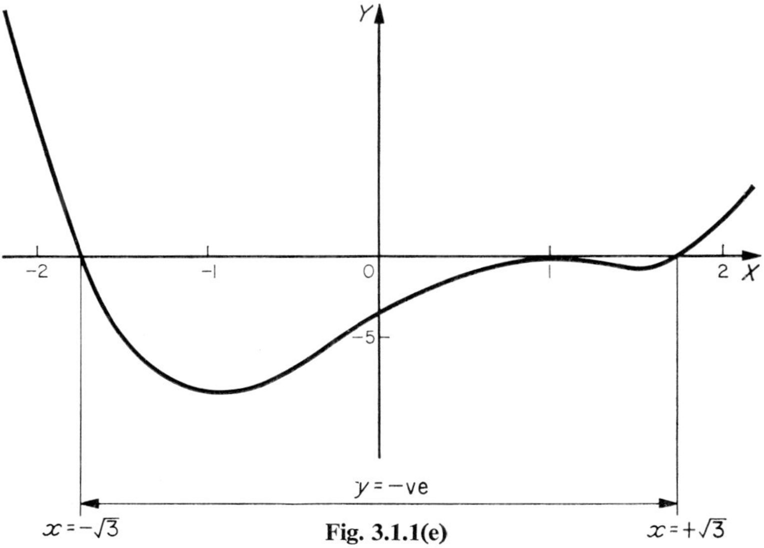

Fig. 3.1.1(e)

Worked example 3

$y = x^2(x - 1)^3 = +x^5 \ldots\ldots\ldots\ldots$

when $y = 0$, $x = 0, 0, 1, 1, 1$ and when $x = 0$, $y = 0$

$(x - 1)^3$ is +ve for $1 < x < +\infty$ and −ve for $-\infty < x < 1$

$\therefore$ y is +ve for $x > 1$ and y is −ve for $x < 1$.

The coefficient of x^5 is +ve $\therefore$ $y \to +\infty$ as $x \to +\infty$

and $y \to -\infty$ as $x \to -\infty$

$\therefore$ The graph is of the form as shown in Fig. 3.1.1(f).

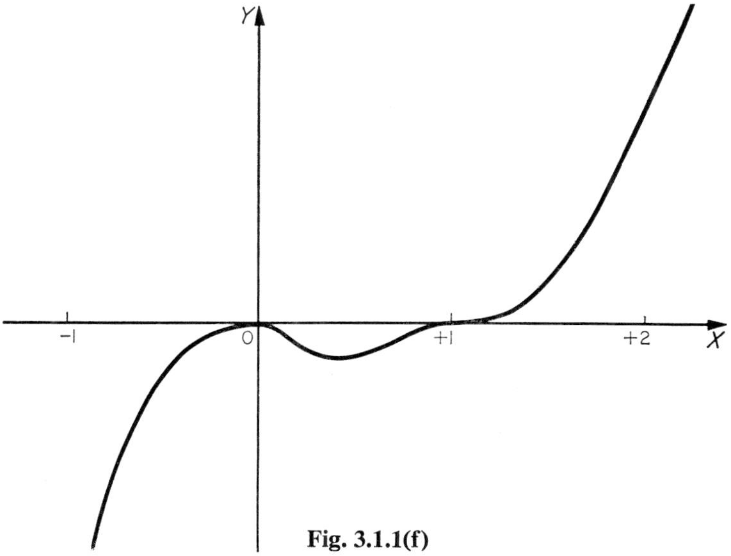

Fig. 3.1.1(f)

For a method of determining the shape of this curve near $x = +1$ see § 3.5.

Worked example 4

$y = 2x^5 - 15x^4 + 30x^3 \equiv x^3(2x^2 - 15x + 30).$

(i) $y = 0$ only when $x = 0$ because the equation $2x^2 - 15x + 30 = 0$ has no real roots.

(ii) $y \to +\infty$ as $x \to +\infty$ and $y \to -\infty$ as $x \to -\infty$

(iii) $(dy/dx) = 10x^4 - 60x^3 + 90x^2 = 10x^2(x - 3)^2$

$\therefore (dy/dx) = 0$ when $x = 0, 3$ and is never negative.

Hence the graph is of the form as shown in Fig. 3.1.1(g).

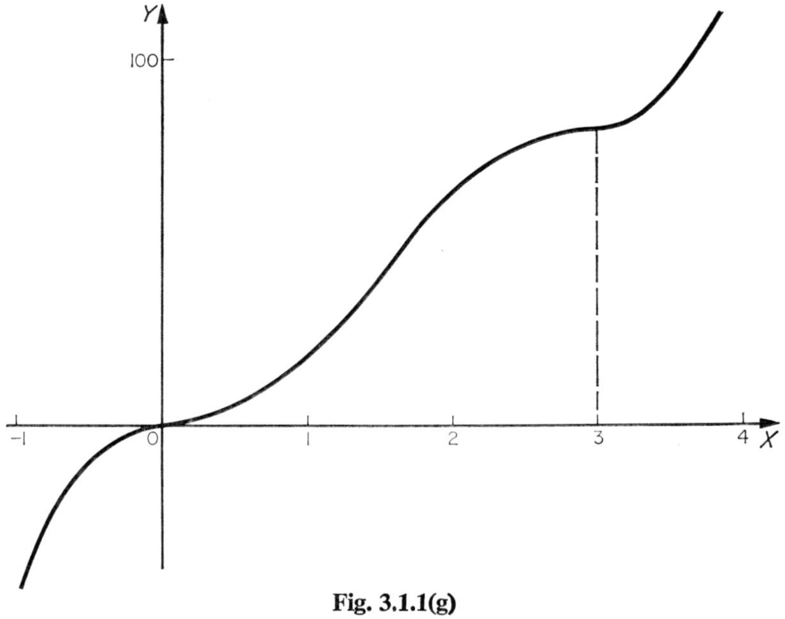

Fig. 3.1.1(g)

3.1.2 Trigonometrical functions (or circular functions)

A line OP, of constant length, is rotated about 0 in the anticlockwise direction starting from its initial position along the axis OX. PN is drawn from P perpendicular to OX.

As the angle XOP increases from zero, NP increases from zero to a maximum (+OP) at the 90° position then decreases to zero at 180°. In the third quadrant NP becomes negative, reaches a minimum value (−OP) at 270° and during the fourth quadrant NP returns to zero at 360°. As the angle XOP increases beyond

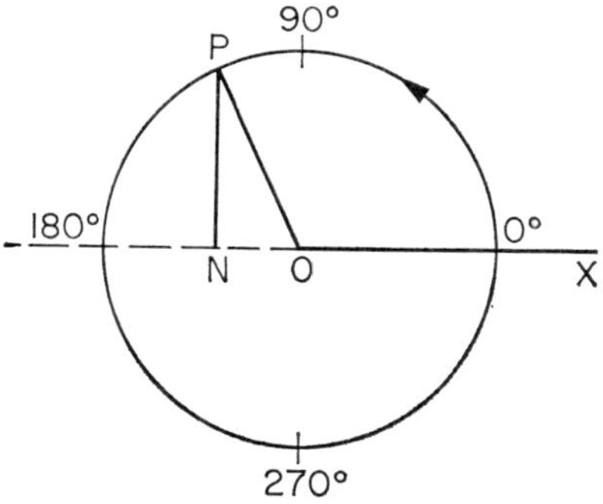

Fig. 3.1.2(a)

360°, the value of NP will repeat the same cycle every time the line OP rotates through a complete turn. During all this movement the length of OP remains constant and is always positive.

Similarly ON varies as the angle increases from 0° to 360° and goes through the following sequence

(+OP) →	0 →	(−OP) →	0 →	(+OP)
at 0°	at 90°	at 180°	at 270°	at 360°

The sine of an angle

$$\text{Definition Sin } X\hat{O}P = (NP/OP)$$

Graph of sin x where x may take all values.

In the Fig. 3.1.2(b) we make OP of unit length, and in this case the value of $\sin x = NP$, numerically.

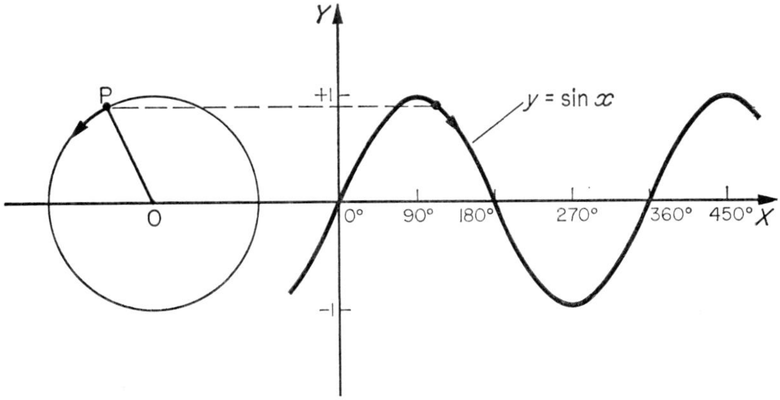

Fig. 3.1.2(b)

The ordinate at any point of the graph is equal to the height of the moving point P above the horizontal diameter of the circle when OP has turned through the appropriate angle.

The cosine of an angle

Definition $\cos X\hat{O}P = (ON/OP)$ (See Fig. 3.1.2(a)).

Graph of $\cos x$ for all values of x, is shown in Fig. 3.1.2(c)

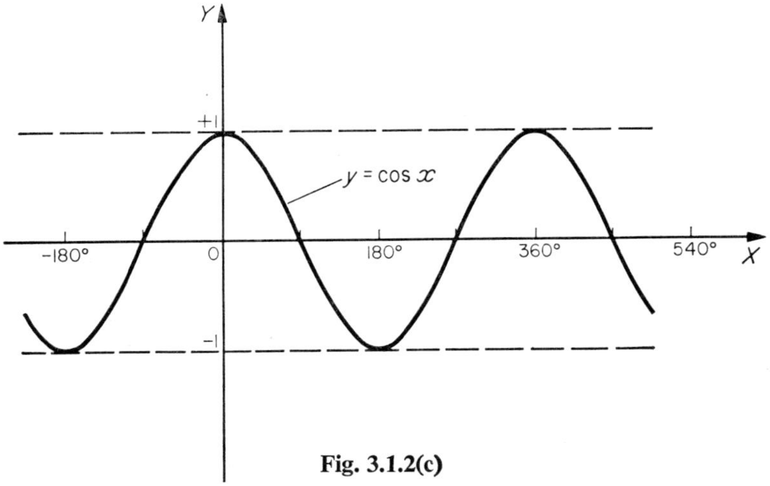

Fig. 3.1.2(c)

With OP again of unit length the curve is drawn so that each ordinate is equal to the length of ON in the appropriate position.

The tangent of an angle

Definition $\tan X\hat{O}P = (NP/ON)$ (See Fig. 3.1.2(a))

Graph of $\tan x$ for all values of x.

In order to follow the changes in the value of $\tan x$ as the angle x increases, it is most convenient to keep the denominator constant (and of unit length).

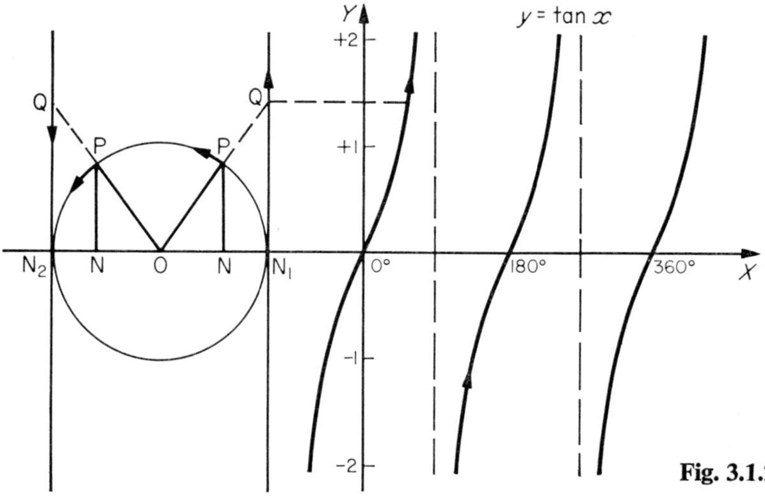

Fig. 3.1.2(d)

In the diagram, for each value of the angle x (i.e. angle NOP) in the *first* and *fourth* quadrants we have:

$$\tan x = \frac{NP}{ON} \text{ (by definition)}$$

$$= \frac{N_1Q}{ON_1} \text{ (by similar triangles)}$$

$$= N_1Q \; (\because ON_1 = +1)$$

This part of the graph can thus be drawn so that each ordinate represents the height of Q for each position of the rotating line OP.

In the *second* and *third* quadrants ($90° < x < 270°$) we have:

$$\tan x = \frac{NP}{ON} \text{ (definition)}$$

$$= \frac{N_2Q}{ON_2} \text{ (by similar triangles)}$$

$$= -N_2Q \; (\because ON_2 = -1)$$

Hence, (allowing for this change of sign) the tangent of the angle x is $-$ve in the second quadrant ($90° < x < 180°$) and $+$ve in the third quadrant ($180° < x < 270°$) but $\tan x$ is still numerically equal to the height of Q above N_1ON_2. The complete graph appears as shown in figure 3.1.2(d) above.

Units of angle—radians

Angles are expressed in several different units. e.g. 60 degrees $\equiv \frac{2}{3}$ of 1 right angle $\equiv \frac{1}{6}$ of 1 complete turn.

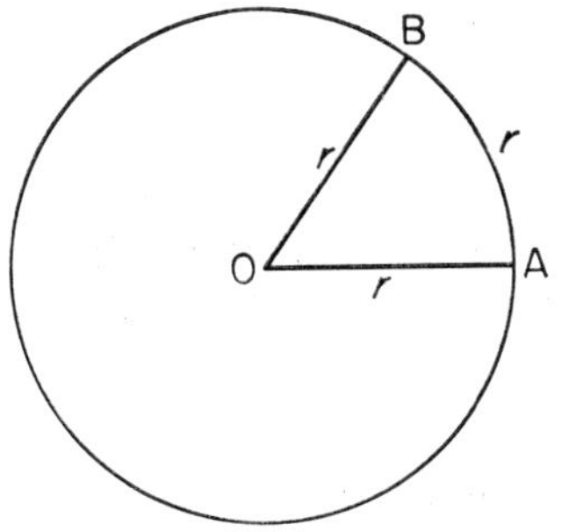

Fig. 3.1.2(e)

Radians—definition. In the diagram, the *arc* AB is of length r, i.e. is equal in length to the radius of the circle. The angle AOB subtended at the centre by this arc is called *one radian.*

The circumference of the circle $= 2\pi r$ (definition of π)

$\therefore$ 2π radians $=$ 1 complete turn $=$ 360 degrees

1 radian $= 180/\pi$ degrees $\simeq 57° 17' 44{\cdot}8''$

1 degree $= \pi/180$ radians $\simeq 0{\cdot}017\,453\,3$ radians

The definition of a radian shows that the magnitude of an angle is a purely numerical quantity. Any angle θ radians subtended at the centre by an arc of a circle is given by $\theta = \dfrac{\text{length of arc}}{\text{radius}} = \dfrac{[L]}{[L]}$ which is dimensionless. In an equation like $2 \sin x = x^2$ the x used must be a number and $\sin x$ is interpreted as the sine of an angle of x radians unless some other unit of angle is specified.

In graphical work we frequently need to use a scale to represent values of an angle, and to have this scale graduated in radians and also in multiples of π radians as shown in the diagram:

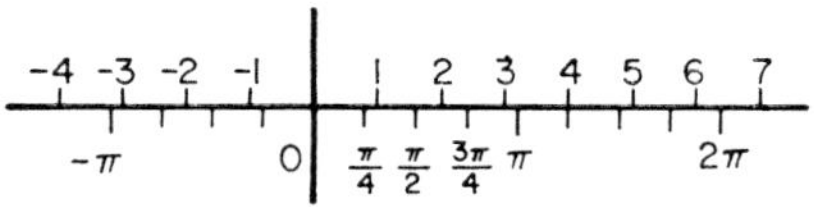

Fig. 3.1.2(f)

The cosecant, secant and contangent of an angle

Definitions: $\operatorname{cosec} x \equiv \dfrac{1}{\sin x}$; $\sec x \equiv \dfrac{1}{\cos x}$;

$$\cot x \equiv \frac{1}{\tan x} \equiv \frac{\cos x}{\sin x}.$$

Graph of cosec x

To sketch the graph of $y = \operatorname{cosec} x$ we start with the graph of $y = \sin x$ and find the ordinate at a number of convenient points.

when $\sin x \to 0$, $\operatorname{cosec} x \to \infty$

as $\sin x$ increases, $\operatorname{cosec} x$ decreases

when $\sin x \to 1$, $\operatorname{cosec} x \to 1$

when $\sin x$ is $-$ve, $\operatorname{cosec} x$ is also $-$ve.

The form of the graph is therefore as shown in Fig. 3.1.2(g).

In the same way, graphs of $\sec x$ and $\cot x$ may be sketched for exercise.

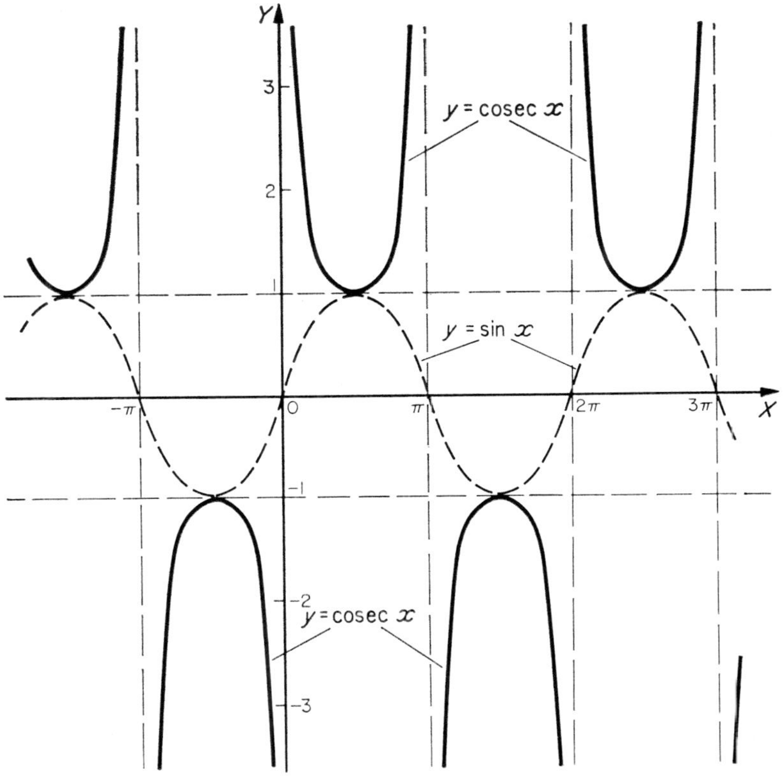

Fig. 3.1.2(g)

Periodic functions

We have seen that as the value of x increases through every range of 2π radians, sin x assumes the same repeated cycle of values. In other words, for all values of x we have $\sin(2\pi + x) = \sin x$. Similarly we have $\cos(2\pi + x) = \cos x$. We describe this property of these functions by saying that sin x and cos x are periodic functions of x, whose period is 2π radians. The values of tan x repeat after π radians, i.e. tan x is a periodic function of x and the period is π radians.

Functions of multiple angles also have their own periods:

e.g. $\cos 2x$ has period π radians (180°)

$\sin 4x$ has period $\pi/2$ radians (90°)

$\sin(x/3)$ has period 6π radians (1080°)

$\tan 3x/4$ has period $4\pi/3$ radians (240°)

The following may be sketched for exercise:

(a) $\sin 5x$ for $-\pi < x < +\pi$

(b) $\cos(x/2)$ for $-3\pi < x < +3\pi$

(c) $\tan(\pi x/4)$ for $-5 < x < +5$

(d) $10\cos 4x$ for $-\pi < x < +\pi$

(e) $5\sin\{3x + (\pi/4)\}$ for $-\pi < x < +\pi$

3.1.3 Exponential functions

In exponential functions, such as 2^x, 10^x and e^{-x^2} the variable occurs in the index. (Note. Indices used to be called exponents).

Graph of $y = a^x$ where a is a constant and $a > 1$ and only positive values of a^x are considered. (See Fig. 3.1.3)

as $x \to -\infty$ $a^x \to 0$
when $x = 0$ $a^x = 1$
as $x \to +\infty$ $a^x \to +\infty$

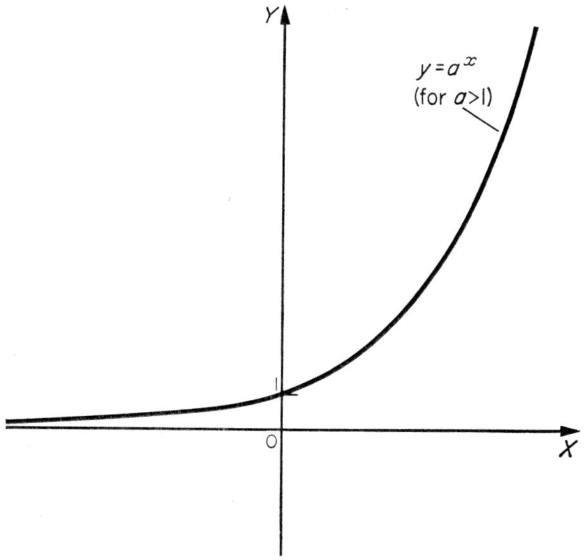

Fig. 3.1.3

as x increases from $-\infty$ to $+\infty$ a^x increases continuously from 0 to $+\infty$, and its characteristic feature is that the rate at which it increases is proportional to its own magnitude (i.e. $dy/dx \propto y$ for all values of x).

(Note: The X axis is the only asymptote)

3.1.4 Logarithmic functions

Definition: If $N = a^n$ then n is called the logarithm of N to the base a i.e. $n = \log_a N$.

Graph of $y = \log_a x$ (where a is a constant and $a > 1$, see note below Fig. 3.1.4)

$$\text{as } x \to +\infty \quad y \to +\infty$$

$$\text{when } x = 1 \quad y = 0$$

$$\text{as } x \to 0 \quad y \to -\infty$$

there are no real values of y when x is negative. As x increases from 0 to $+\infty$ y increases from $-\infty$ to $+\infty$. The rate at which y increases is inversely proportional to the value of x. (i.e. $dy/dx \propto 1/x$).

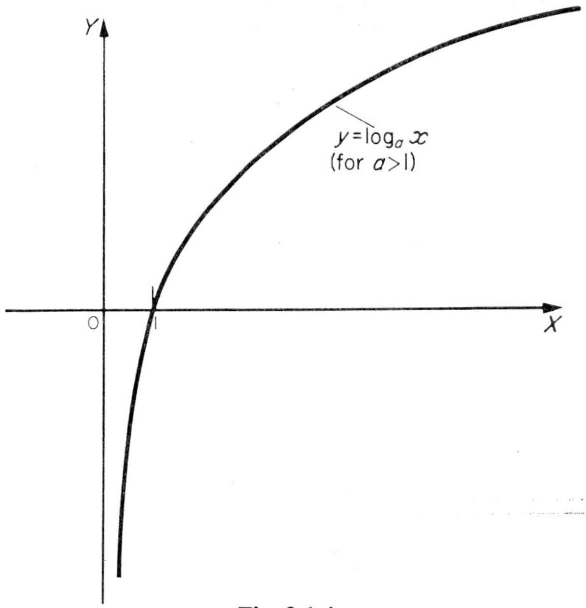

Fig. 3.1.4

Note. The graphs of a^x and $\log_a x$ consist of two similar curves. If one of them is rotated through two right angles about the straight line $y = x$ it will coincide with the other. Each curve is the reflection of the other in the line $y = x$.

By compiling a table of powers of 2 the following graphs may be sketched for exercise:

1. $y = 2^x$
2. $y = 2^{-x}$
3. $y = \log_2 x$
4. $y = \log_{\frac{1}{2}} x$
5. $y = 2^x + 2^{-x}$
6. $y = 2^x - 2^{-x}$

3.1.5 Hyperbolic functions

Six Hyperbolic functions are formed by combining the exponential functions e^x and e^{-x}. They have names which correspond with the names of the six trigonometrical functions, because the two groups of functions have many similarities. These resemblances do not appear in the graphs of the functions, but soon become clear when formulae and identities are considered.

Hyperbolic sine and cosine

Definitions: $\sinh x \equiv \frac{1}{2}(e^x - e^{-x})$

$$\cosh x \equiv \frac{1}{2}(e^x + e^{-x})$$

From these definitions it is easy to show that $\cosh^2 x - \sinh^2 x \equiv 1$ for all values of x. This identity compares with the corresponding identity $\cos^2 x + \sin^2 x \equiv 1$ for trigonometrical functions.

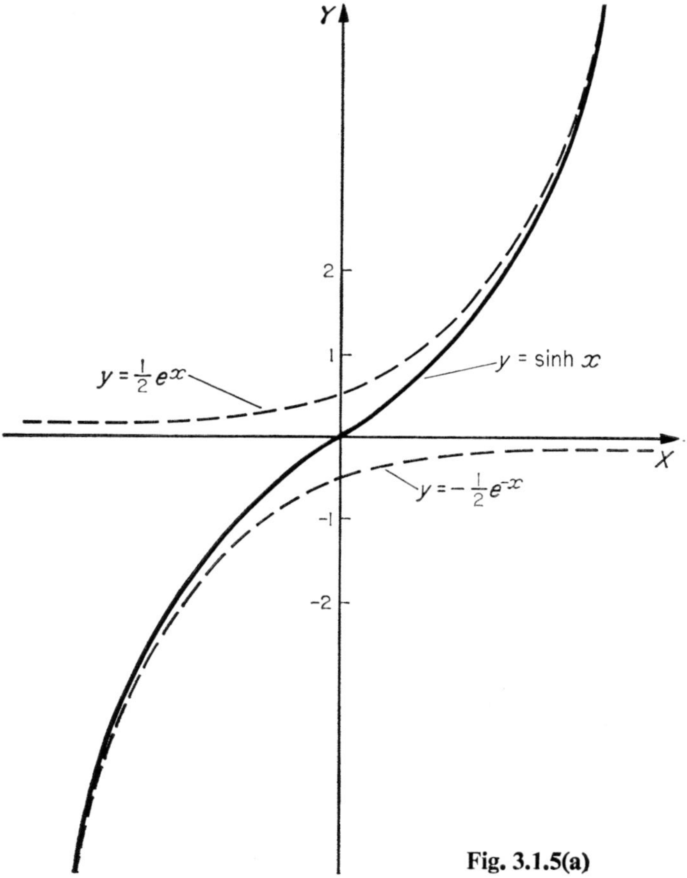

Fig. 3.1.5(a)

Graph of $y = \sinh x \equiv \frac{1}{2}(e^x - e^{-x})$

as $x \to +\infty$, $e^x \to +\infty$ and $e^{-x} \to 0$

$\therefore \sinh x < \frac{1}{2}e^x$ and $\sinh x \to \frac{1}{2}e^x \to +\infty$

when $x = 0$ $\sinh x = 0$

as $x \to -\infty$, $e^x \to 0$ and $e^{-x} \to +\infty$

$\therefore \sinh x \to -\frac{1}{2}e^{-x} \to -\infty$

Fig. 3.1.5(a) shows the graphs of $y = \frac{1}{2}e^x$, $y = -\frac{1}{2}e^{-x}$, and $y = \sinh x$. Graph of $y = \cosh x \equiv \frac{1}{2}(e^x + e^{-x})$ is shown below in Fig. 3.1.5(b)

as $x \to +\infty$ $\quad y \to \frac{1}{2}e^x \to +\infty$

when $x = 0$ $\quad y = 1$

as $x \to -\infty$ $\quad y \to \frac{1}{2}e^{-x} \to +\infty$

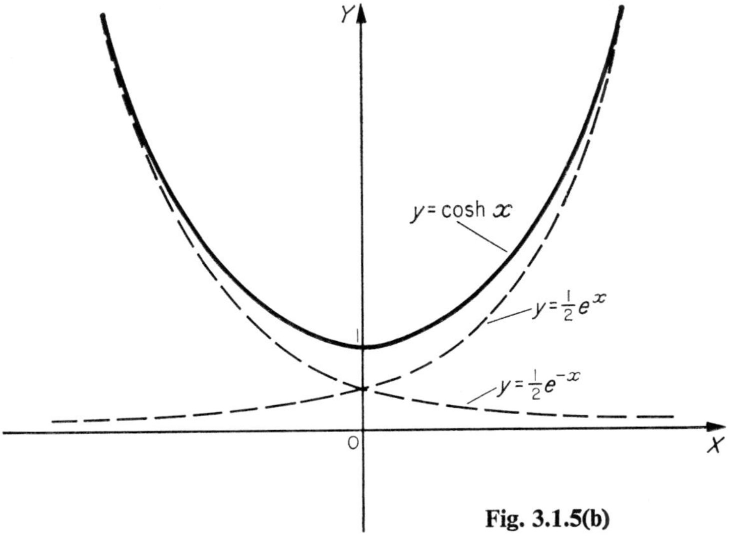

Fig. 3.1.5(b)

Hyperbolic tangent

Definition: $\tanh x = \dfrac{\sinh x}{\cosh x}$ (As in trigonometry)

$$\therefore \tanh x = \frac{e^x - e^{-x}}{e^x + e^{-x}} \quad (1) \quad = \frac{e^{2x} - 1}{e^{2x} + 1} \quad (2) \quad = \frac{1 - e^{-2x}}{1 + e^{-2x}} \quad (3)$$

Graph of $y = \tanh x$ is shown below in Fig. 3.1.5(c)

As $x \to +\infty$ $e^{-2x} \to 0$

$\therefore$ using (3) $\tanh x < 1$ and $\tanh x \to 1$. When $x = 0$ $\tanh x = 0$

As $x \to -\infty$ $e^{2x} \to$;

$\therefore$ using (2) $\tanh x > -1$ and $\tanh x \to -1$.

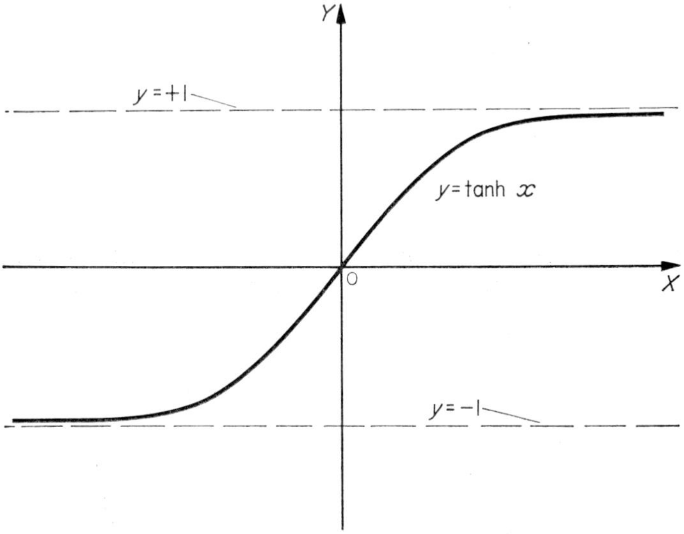

Fig. 3.1.5(c)

Hyperbolic cosecant, secant, and cotangent

Definitions:

$$\text{cosech } x \equiv \frac{1}{\sinh x}$$

$$\text{sech } x \equiv \frac{1}{\cosh x}$$

$$\coth x \equiv \frac{1}{\tanh x}$$

The graphs of $y = \text{cosech } x$, $y = \text{sech } x$ and $y = \coth x$, using definitions and the method outlined for cosec x in § 3.1.2(end), may be sketched for exercise.

3.2 GENERAL CURVE SKETCHING

3.2.1 Investigation of basic features

The most elementary features to examine first when planning a sketch of $y = f(x)$ or $f(x, y) = 0$* are:

1. The values of y when $x = 0$
2. The values of x when $y = 0$
3. The values of y when $x \to \pm\infty$
4. The values of x when $y \to \pm\infty$
5. Any other obvious numerical values.

* The notation $f(x, y)$ refers to a function of x and y, for example $x^2 + y^2 + 2x + 2y - 7$.

Not all of the above information about the curve can be expected to be easily obtainable in all cases, and at this stage all laborious calculations should be avoided.

Worked example 1.

$xy - y - 2x + 1 = 0$ (1) (An example of $f(x,y) = 0$ although in this case it is possible to rearrange the given equation into the form $y = g(x)$)

This can be arranged as $y = \dfrac{2x - 1}{x - 1} = 2 + \dfrac{1}{x - 1}$ (2)

or again as $x = \dfrac{y - 1}{y - 2} = 1 + \dfrac{1}{y - 2}$ (3)

we have 1. $x = 0$ only when $y = 1$

2. $y = 0$ only when $x = \frac{1}{2}$

3. as $x \to +\infty$ $y > +2$ and $y \to +2$ from (2)
as $x \to -\infty$ $y < +2$ and $y \to +2$ from (2)

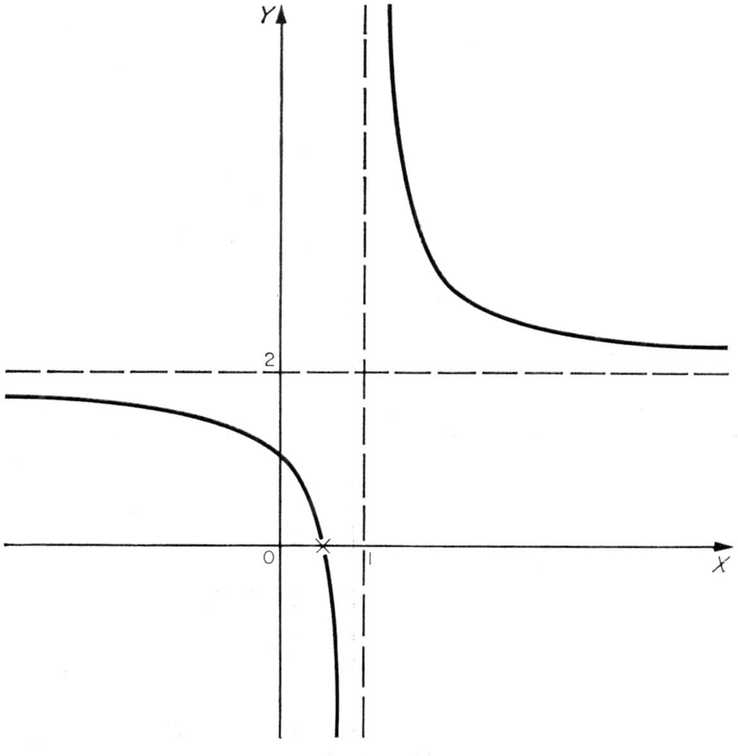

Fig. 3.2.1(a)

4. as $y \to +\infty$ $x > +1$ and $x \to +1$ from (3)
as $y \to -\infty$ $x < +1$ and $x \to +1$ from (3)

5. when $x > 1$ $y > 2$

Hence the graph is of the form shown in Fig. 3.2.1(a)

Worked example 2

$$y = \frac{x+2}{(x-1)(x-3)}$$

Here $y = 0$ only when $x = -2$

$y \to \infty$ when $x \to 1$ and when $x \to 3$

$y \to 0$ when $x \to +\infty$ and when $x \to -\infty$

$y = -$ve in the range $-\infty < x < -2$ (All three factors are negative)

$y = +$ve in the range $-2 < x < 1$ (Two factors are $-$ve)

$y = -$ve in the range $1 < x < 3$ (One factor is $-$ve)

$y = +$ve in the range $3 < x < \infty$ (All factors $+$ve).

Hence the graph is of the form shown in Fig. 3.2.1(b)

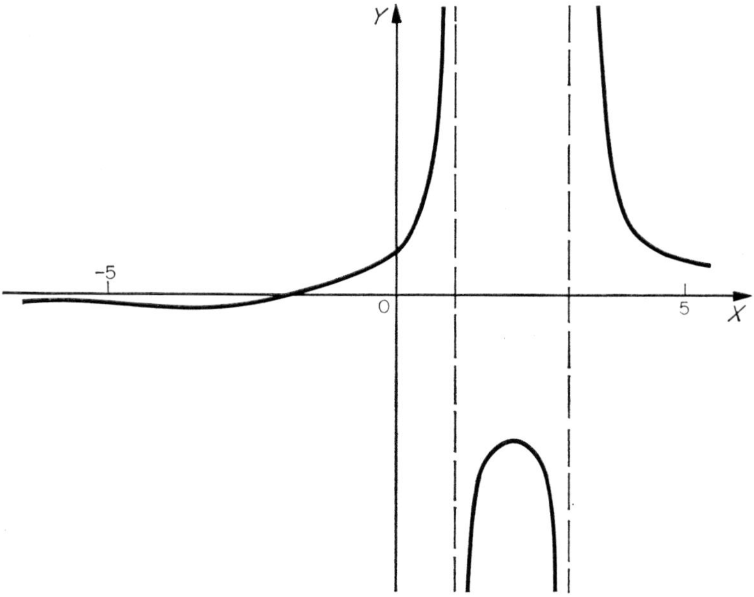

Fig. 3.2.1(b)

3.2.2 Symmetry

(a) Functions such as cos x, sec x, cosh x, sech x, x^2, $x^2 - x^6$, etc., are called *even functions* of x. They all have the common property that $f(-x) \equiv f(x)$.

e.g. $(e^x - e^{-x})/x$ is an even function. This can be proved as follows:

$$\text{let } f(x) \equiv \frac{e^x - e^{-x}}{x}$$

$$\text{then } f(-x) \equiv \frac{e^{-x} - e^{+x}}{-x} \equiv \frac{e^x - e^{-x}}{x} \quad \text{which is identical to } f(x).$$

This property $f(x) \equiv f(-x)$ shows that the graph of an even function of x is symmetrical about OY as in Fig. 3.2.2(a) below.

(b) Functions such as sin x, tan x, sinh x, tanh x, x^3, $x^3 - x^5$, etc., are called *odd functions* of x. They all have the property that $f(-x) \equiv -f(+x)$.

e.g. $(x^2 - x \sin x)/(\tan x)$ is an odd function, since it retains its magnitude but changes its sign when $-x$ is substituted for x.

This property $f(-x) \equiv -f(+x)$ implies that the graph of an odd function of x has a special property and this is illustrated in Fig. 3.2.2(b) below.

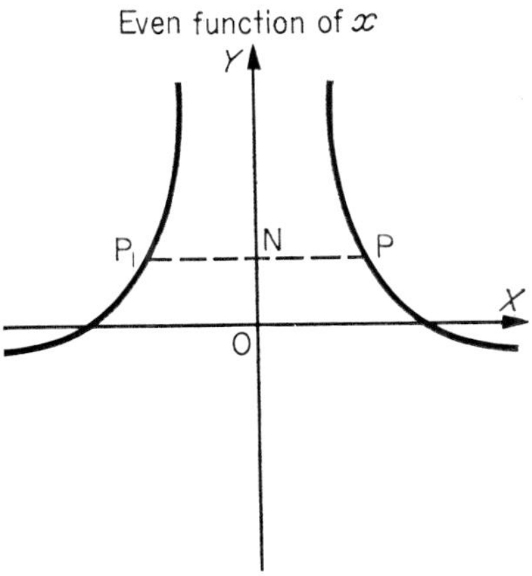

Fig. 3.2.2(a)

$f(-a) \equiv f(+a)$

and $\therefore$ $P_1N = NP$

Reflexion in OY causes one half to coincide with the other.

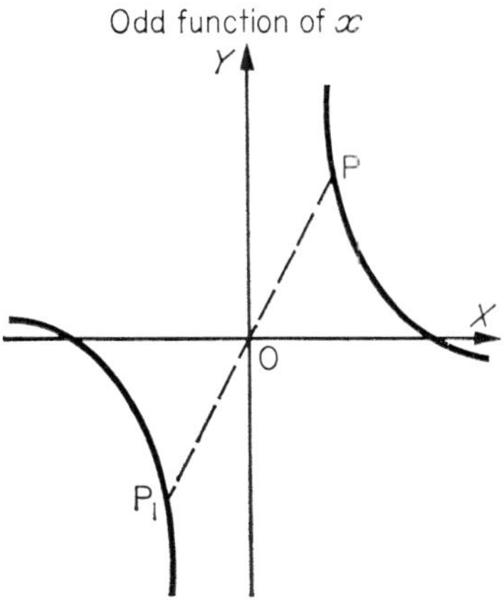

Fig. 3.2.2(b)

$f(-a) \equiv -f(+a)$

and $\therefore$ $P_10 = OP$

Rotation about the origin as a centre causes one half to coincide with the other.

Worked example 1

$y = x^4 - 5x^2 + 4$ is an even function of x, see worked example in § 3.1.1.

Worked Example 2

$y = x/(1 + x^2)$ is an odd function of x.

when $x = 0, y = 0$, when x is $-$ve, y is $-$ve.

as $x \to \pm \infty$ $y \to 0$.

if x is small $y \simeq x$ $\therefore$ gradient at $0 = 1$

$\therefore$ the curve is as shown in Fig. 3.2.2(c).

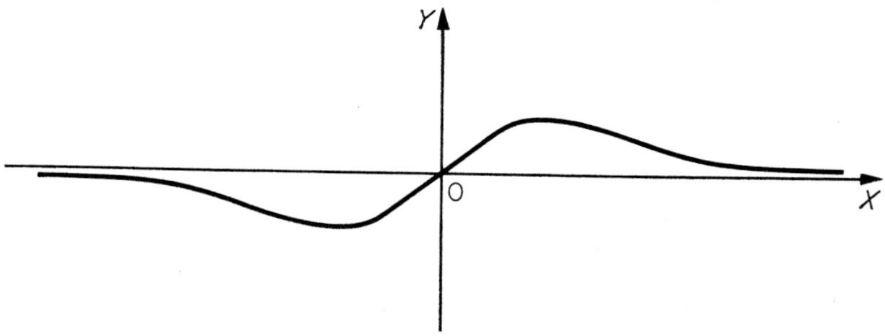

Fig. 3.2.2(c)

Worked example 3

$$y^2 = x^2(2 - x^2) \quad \text{. . . . . (1)}$$

This can be arranged as $x^4 = 2x^2 - y^2$ . . . . (2)

or again as $x^2 = 1 \pm \sqrt{(1 - y^2)}$. . . . . (3)

We have: (a) The curve is symmetrical about both OX and OY.

(b) $x = 0$ only when $y = 0$

(c) $y = 0$ only when $x = -\sqrt{2}, 0, +\sqrt{2}$

(d) y is real only when $x^2 \leqslant 2$ from (1)

(e) x is real only when $y^2 \leqslant 1$ from (3).

Hence no part of the curve lies outside the rectangle formed by the lines $y = \pm 1, x = \pm\sqrt{2}$.

Again from equation (2) x is real only when $y^2 \leqslant 2x^2$ i.e. when y is between $-\sqrt{(2x)}$ and $+\sqrt{(2x)}$. $\therefore$ No part of the curve is outside the area shown in Fig. 3.2.2(d).

This information is sufficient to indicate the figure eight shape of the curve. More detailed examination of the curve near O and A would be necessary to complete the sketch. (See § 3.4 for method).

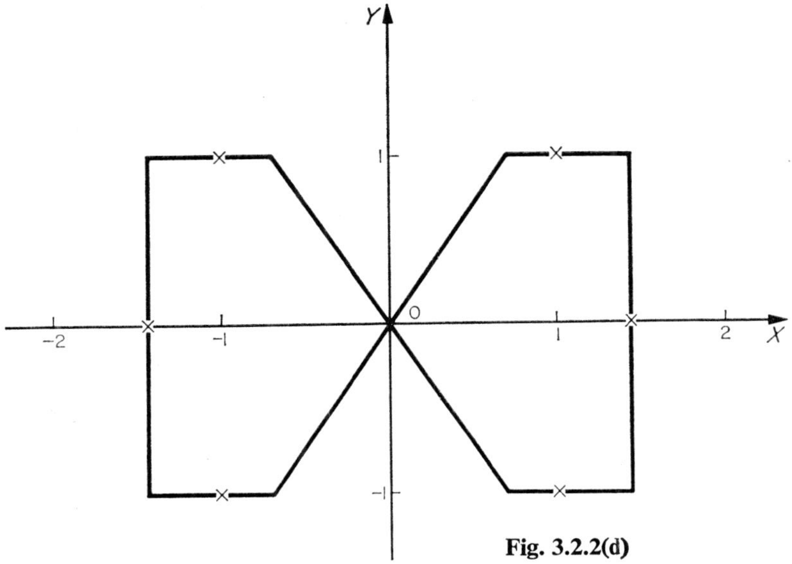

Fig. 3.2.2(d)

3.3 CURVES OBTAINED BY COMBINING SIMPLER CURVES

3.3.1 Addition or subtraction of ordinates

Example 1

$y = (x^3)/(1 + x^2)$ this is $= x - x/(1 + x^2)$

Method: Sketch $y_1 = y_2 = x$ and $x/(1 + x^2)$

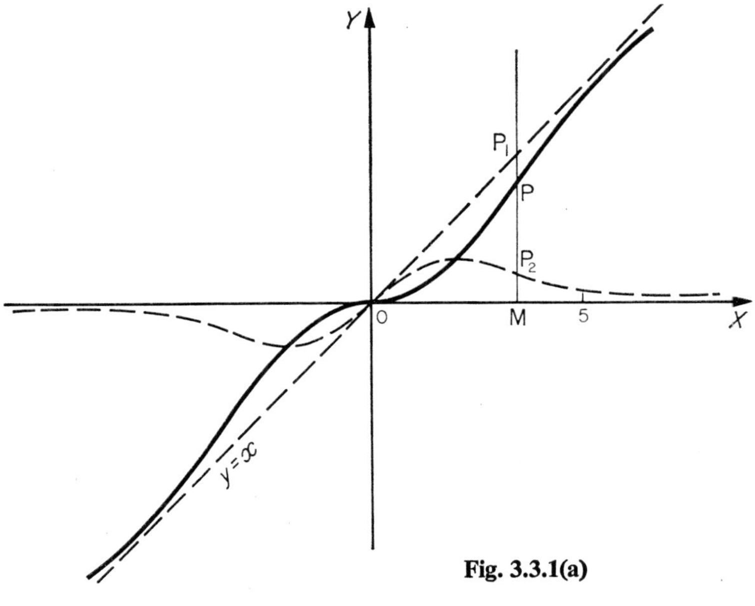

Fig. 3.3.1(a)

The points (P) on the required curve are obtained by making $P_1P = P_2M$· i.e. $PM = y_1 - y_2$ everywhere.

Example 2

$y = \sin 3x + 3 \sin x.$

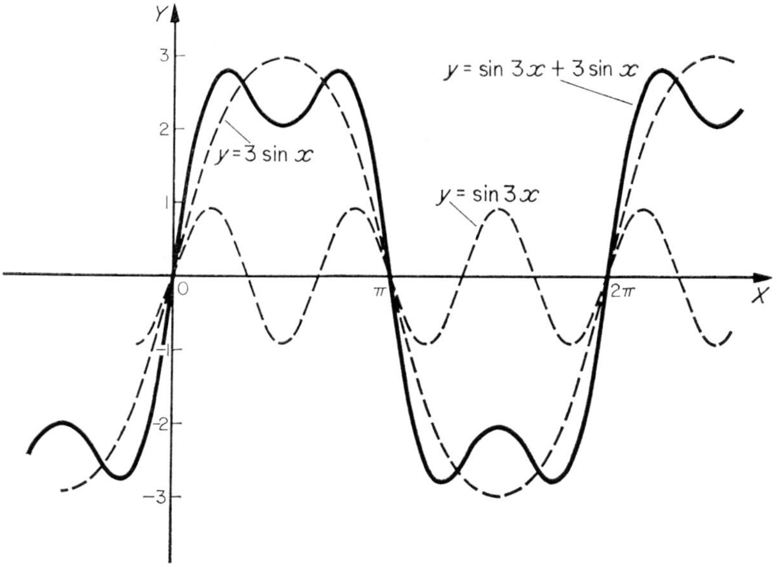

Fig. 3.3.1(b)

3.3.2 Multiplication of ordinates

$y = x \sin x.$

Draw $y_1 = x$ and $y_2 = \sin x$ on the same diagram.

At each point such as P on the required curve $PM = P_1M \times P_2M$ i.e. $y = y_1 \times y_2$.

It is useful to graduate the X axis in integers and also in multiples of π.

Notice the tangency at A where $x = \pi/2$

and at B where $x = 3\pi/2$

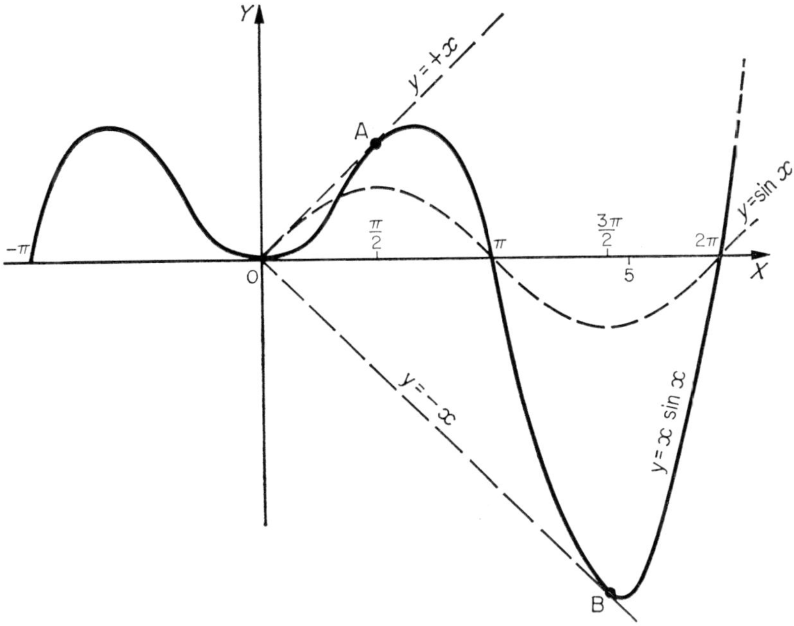

Fig. 3.3.2

3.3.3 Square roots of ordinates

In order to sketch the curve $y^2 = \dfrac{2x + 3}{4(x - 1)}$ first sketch $y_1 = \dfrac{2x + 3}{4(x - 1)}$

At each point such as P on the required curve $MP = \pm\sqrt{(MP_1)} = \pm\sqrt{(y_1)}$

Note: where $y_1 > 1$ $\quad |y| < y_1$

where $0 < y_1 < 1$ $\quad |y| > y_1$

and where y_1 is $-$ve $\quad y$ does not exist.

As $x \to \pm\infty$ $y_1 \to +\frac{1}{2}$ and $y \to \pm\dfrac{1}{\sqrt{2}} = \pm 0{\cdot}707$

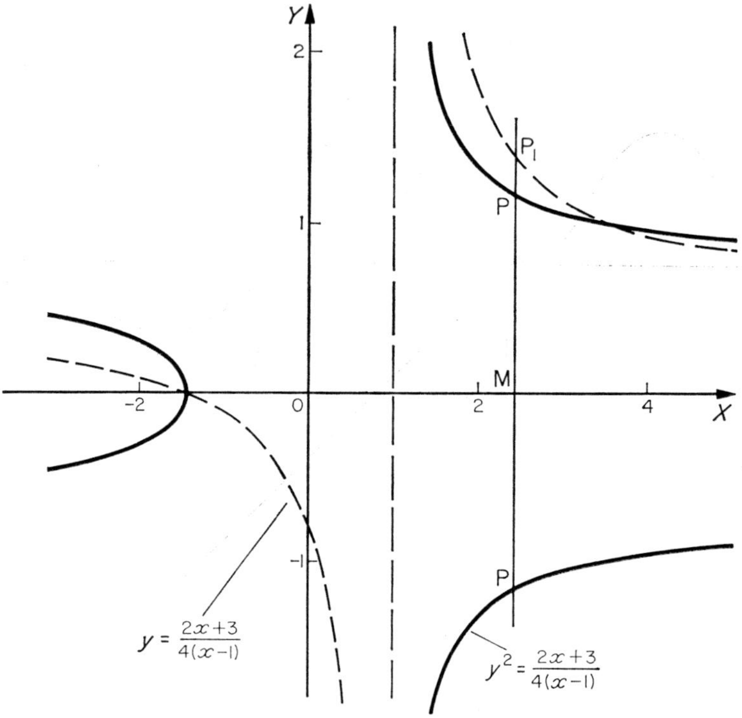

Fig. 3.3.3

3.4 APPROXIMATIONS TO THE SHAPE OF A CURVE NEAR SPECIAL POINTS

(a) $y = \dfrac{2x+3}{4(x-1)}$ *near* $x = 1$

let $x = 1 + h$ where h is small and can be either positive or negative

then $y = \dfrac{5+2h}{4h}$ which $\rightarrow \dfrac{5}{4h}$ as $h \rightarrow 0$ because $2h$ is small compared with 5.

$\therefore$ when h is small and +ve y is large and +ve and when h is small and −ve y is large and −ve.

Hence the shape of the curve near the asymptote $x = 1$ is as shown in Figure 3.4(a).

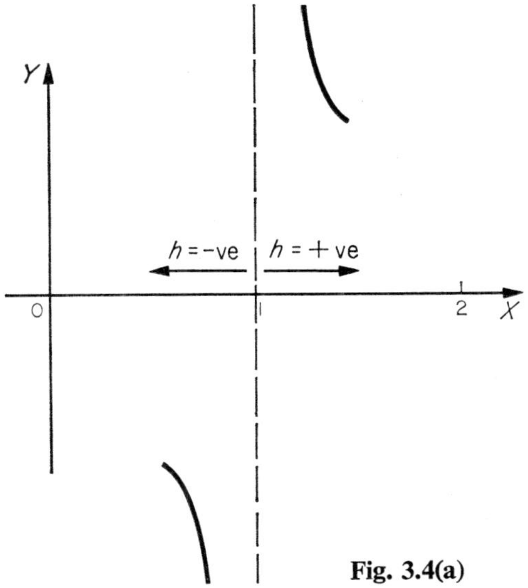

Fig. 3.4(a)

(b) $y^2 = \dfrac{2x + 3}{4(x - 1)}$ *near* $x = -1{\cdot}5$

let $x = -1{\cdot}5 + h$

then $y^2 = \dfrac{2h}{-10 + 4h}$ i.e. $y^2 \rightarrow -\frac{1}{5}h$ as $h \rightarrow 0$ because $4h$ is small compared with 10.

$\therefore$ there are no real values of y when h is positive i.e. when $x > -1{\cdot}5$

and when h is negative (i.e. $x < -1{\cdot}5$) y is small and the curve approximates to the parabola $y^2 = -\frac{1}{5}$h.

Hence the shape of the curve near the point $(-1{\cdot}5, 0)$ is approximately the parabola shown in Figure 3.4(b).

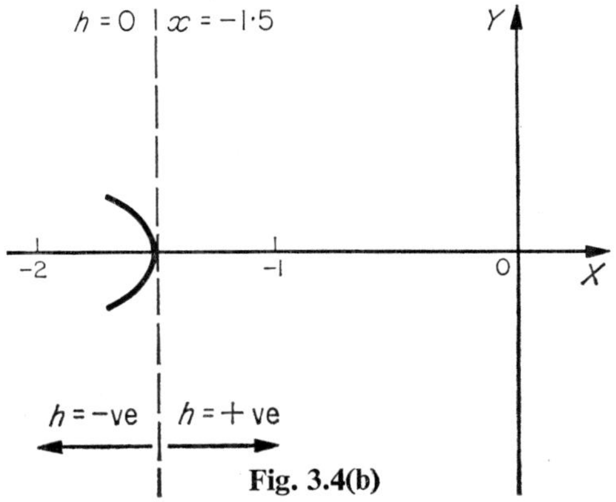

Fig. 3.4(b)

(c) $y = x \sin x$ *near* $x = \pi/2$

let $x = \pi/2 + h$

then $y = \left(\frac{\pi}{2} + h\right)\cos h$ i.e. $y \simeq \frac{\pi}{2} + h.$

i.e. for values of x close to $\pi/2$ the equation of the curve approximates to $y = x$ and hence the curve passes through the point A, $(\pi/2, \pi/2)$ with gradient 1 (See § 3.3.2).

(d) Sketch $y = \dfrac{x^3 + 3x^2 - 3x - 10}{x^2 - 3}$ . . . (i)

$$\equiv \frac{(x + 2)(x^2 + x - 5)}{[x + \sqrt{(3)}][x - \sqrt{(3)}]} \quad \ldots \text{(ii)}$$

$$\equiv x + 3 - \frac{1}{x^2 - 3} \quad \ldots \text{(iii)}$$

when $y = 0$ x $= -2, -2{\cdot}79 + 1{\cdot}79$

when $x = 0$ $y = 3\frac{1}{3}$

near $x = \sqrt{(3)}$

put $x = \sqrt{(3)} + h$ in (ii) giving

$$y \simeq \frac{[\sqrt{(3)} + 2][\sqrt{(3)} - 2]}{2\sqrt{(3)}\,.\,h} \to -\frac{1}{2\sqrt{(3)}\,h}$$

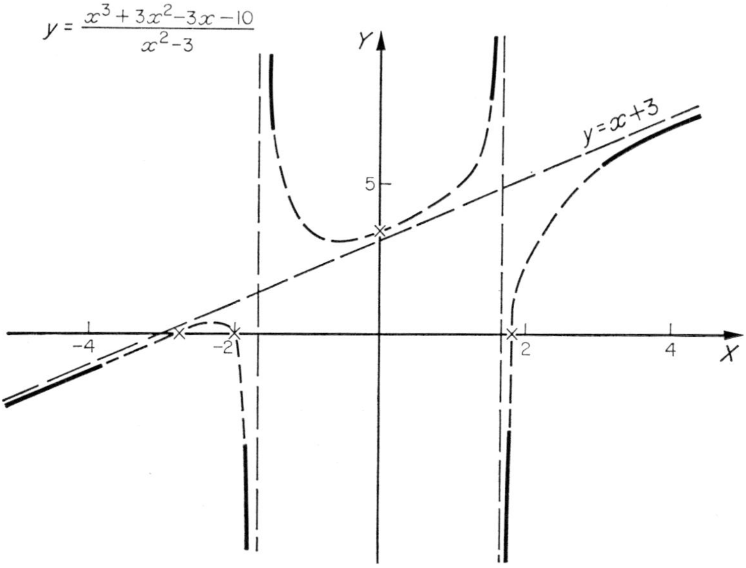

Fig. 3.4(d)

near $x = -\sqrt{(3)}$ put $x = -\sqrt{(3)} + h$ giving $y \simeq +\dfrac{1}{2\sqrt{(3)}\,h}$

near $x = 0$ $y \simeq 3\frac{1}{3} + x + \ldots$ from (iii)

as $x \to \pm\infty$ $y \to x + 3$ and also $y < x + 3$ $\therefore$ $y = x + 3$ is an asymptote.

These results are shown in figure 3.4 (d) and the dotted lines indicate how the rest of the curve can be sketched in.

3.5 USE OF CALCULUS

Stationary values and points of inflexion are sometimes of great assistance in sketching a curve. For example: sketch $y = e^{-x^2}$

$$\frac{dy}{dx} = -2x\,e^{-x^2}$$

$$\frac{d^2y}{dx^2} = 4x^2\,e^{-x^2} - 2e^{-x^2} = 0 \text{ when } x = \pm\frac{1}{\sqrt{(2)}}$$

There is a point of inflexion at $[1/\sqrt{(2)}, e^{-\frac{1}{2}}]$ and the equation of the tangent at this point is $y - e^{-\frac{1}{2}} = -\sqrt{(2)}\,e^{-\frac{1}{2}}\,[x - (1/\sqrt{2})]$ and this line meets the axes at $(0, 2e^{-\frac{1}{2}})$ and $[\sqrt{2}, 0]$.

when $x = \sqrt{2}$ $\quad y = e^{-2}$

when $x = 0$ $\quad y = 1$.

Plotting these points and drawing the tangent which gives the direction of the curve at its point of inflexion we have:

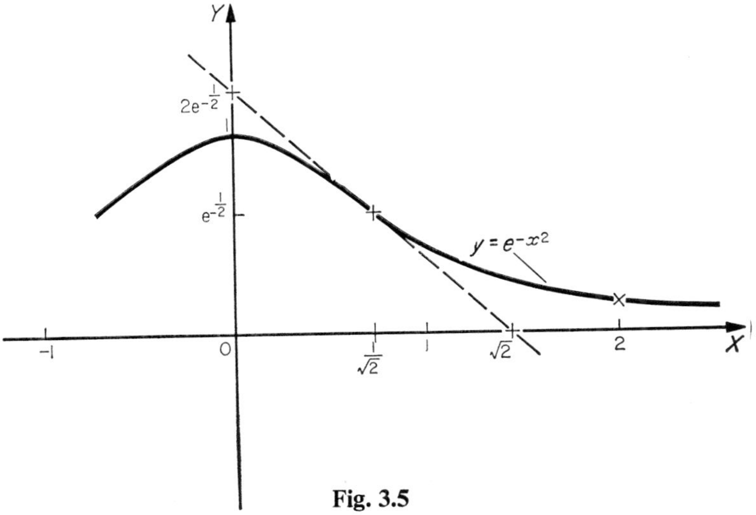

Fig. 3.5

and the whole curve is symmetrical about OY.

3.6 GRAPHICAL SOLUTION OF EQUATIONS

The graph of $y = f(x)$ meets the x axis at values of x which are the roots of the equation $f(x) = 0$. If the equation $f(x) = 0$ can be re-arranged in the form $F_1(x) = F_2(x)$, then the curves $y = F_1(x)$ and $y = F_2(x)$ intersect in points whose abscissae are the roots of the equation $f(x) = 0$. The functions $F_1(x)$ and $F_2(x)$ can often be chosen so that they can be sketched much more easily than $f(x)$.

Worked Example 1

$x^4 + 3x - 1 = 0$ is the same as $x^4 = 1 - 3x$. The diagram shows $y = x^4$ and $y = 1 - 3x$.

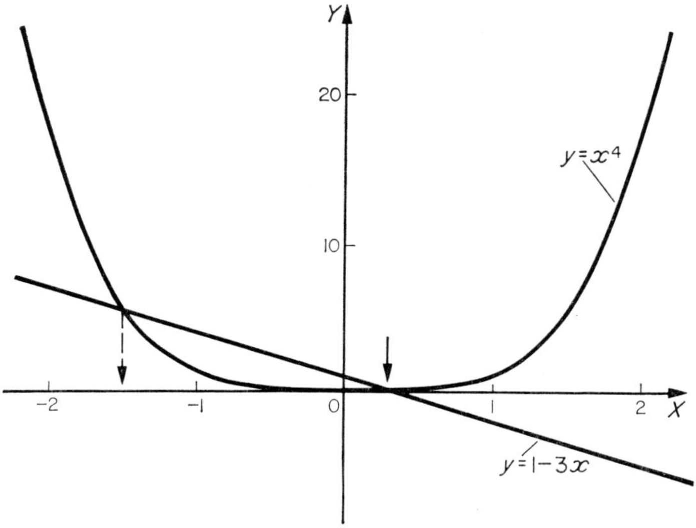

Fig. 3.6(a)

There are only two intersections at approximately $x = -1{\cdot}6$ and $x = +0{\cdot}3$ and therefore the only real roots of the equation $x^4 + 3x - 1 = 0$ are near $-1{\cdot}6$ and $0{\cdot}3$.

Worked example 2

$x \sin x = 1$ or $x = \operatorname{cosec} x$ or $\sin x = 1/x$. The third of these forms is the easiest to sketch.

∴ There is an infinite number of roots of the equation $x \sin x = 1$ whose values are *near* the following:

$$x = \pm 1{\cdot}1, \pm 2{\cdot}9, \pm 6{\cdot}4 \ldots \pm n\pi \ldots$$

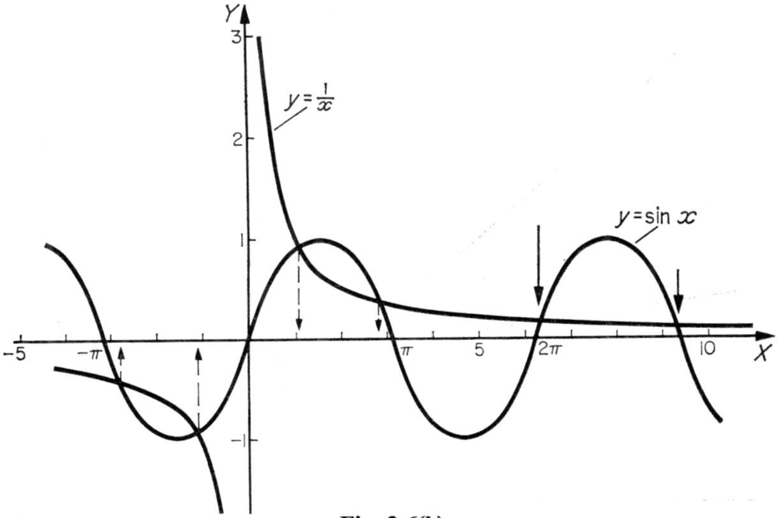

Fig. 3.6(b)

3.6.1 Obtaining better accuracy

Having found approximate values of the roots of an equation, each of these roots can be obtained to greater accuracy by drawing a small part of the graph on a much larger scale. Consider the root of $x^4 + 3x - 1 = 0$ which is near $-1{\cdot}6$. Tabulating the values for x near $-1{\cdot}6$ we have,

x	x^4	$1 - 3x$
$-1{\cdot}5$	$5{\cdot}0625$	$5{\cdot}5$ $(x^4 < 1 - 3x)$
$-1{\cdot}6$	$6{\cdot}5536$	$5{\cdot}8$ $(x^4 > 1 - 3x)$
$-1{\cdot}7$	$8{\cdot}3521$	$6{\cdot}1$

This enlarged portion of the graph shows that a better approximation to the root is $x = -1{\cdot}54$. Notice that, to the left of the root $x^4 > 1 - 3x$ and to the right of the root $x^4 < 1 - 3x$. Also $y = 1 - 3x$ is a straight line, and $y = x^4$ although a curve appears to be very nearly straight when drawn on this scale.

From the above table $P_1Q_1 = 0{\cdot}7536$ and $Q_2P_2 = 0{\cdot}4375$

The Fig. 3.6.1 (b) shows the straight line joining Q_1 and Q_2 and from the pair of similar triangles we deduce that R divides BA in the ratio 0·7536:0·4375

$$\text{and } \therefore \text{ RA} = \frac{4375}{11911} \times 0{\cdot}1$$

$$\simeq \frac{44}{120} \times 0{\cdot}1 = 0{\cdot}037 \text{ (approx.)}$$

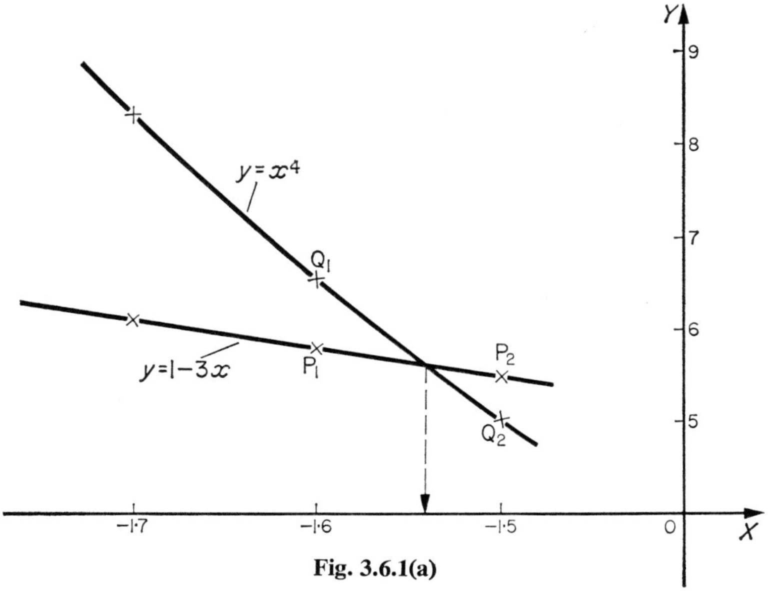

Fig. 3.6.1(a)

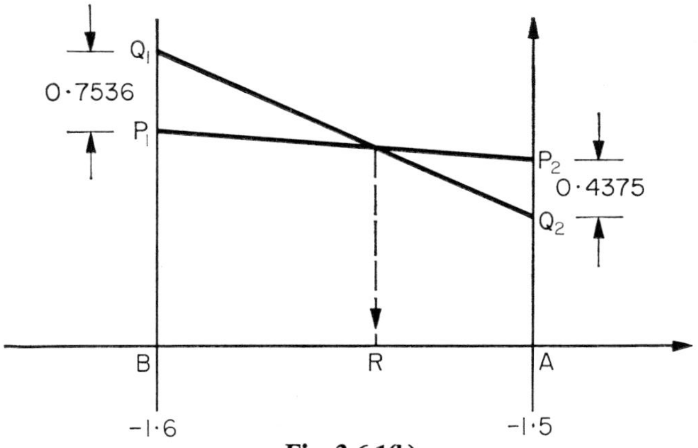

Fig. 3.6.1(b)

At this stage we remember that we have just used a straight line Q_1Q_2 instead of the actual arc Q_1Q_2 of the curve $y = x^4$, and that therefore the correct point of intersection would be slightly to the left of the point R whose position we have just calculated. This shows that the root is slightly to the left of $x = -1{\cdot}537$ and that probably $-1{\cdot}540 < x < -1{\cdot}537$.

Example

Complete the table below only so far as it is necessary (6 figure tables or a calculating machine are required) to find two values of x which 'straddle' the root.

x	x^4	$1 - 3x$	Difference
$-1{\cdot}537$			
$-1{\cdot}538$			
$-1{\cdot}539$			
$-1{\cdot}540$			

Hence, estimate the value of the root to five decimal places.

This simple method of estimating a root of an equation is usually used to find a good first approximation to the root. Later in Chapter 4 we show how to obtain, more systematically, a value of the root which is correct to any required degree of accuracy.

3.6.2 Use of linear interpolation

The procedure used in the above paragraph is called linear interpolation. It can be used whenever the equation $f(x) = 0$ has only one simple root between $x = a$ and $x = b$ where $b - a$ is small. Under these conditions $f(a)$ and $f(b)$ are both small and have opposite signs.

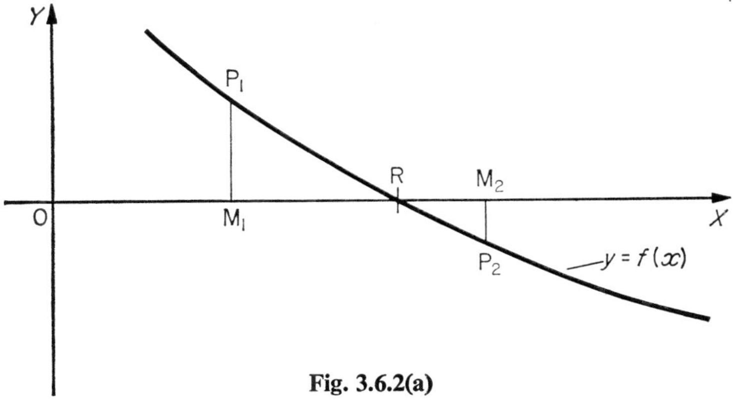

Fig. 3.6.2(a)

In Fig. 3.6.2 (a) $M_1P_1 = f(a)$ is the ordinate at $x = a$

and $M_2P_2 = f(b)$ is the ordinate at $x = b$

The root (R) plainly divides M_1M_2 approximately in the ratio $M_1P_1:P_2M_2$ i.e. $|f(a)| : |f(b)|$

Similarly if $f(x) = 0$ is expressed as $F_1(x) = F_2(x)$ the previous figure becomes Fig. 3.6.2 (b) and the root R divides M_1M_2 approximately in the ratio $Q_1P_1:P_2Q_2$ i.e. in the ratio

$$F_1(a) \sim F_2(a):F_1(b) \sim F_2(b)$$

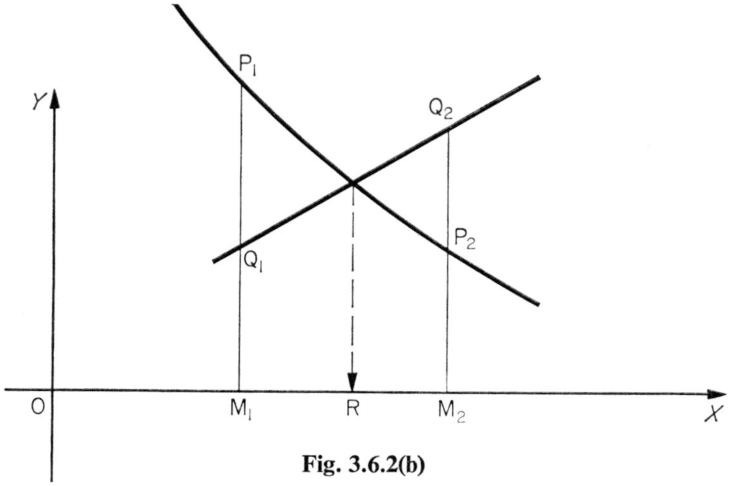

Fig. 3.6.2(b)

In practise this provides a very simple method of improving the accuracy of an estimate of a root of an equation without having to draw a large scale accurate graph.

For example: The sketch in § 3.3.2 shows that the equation $x \sin x = 1$ has one root near $x = 1{\cdot}1$. Tabulating values for $x = 1{\cdot}0, 1{\cdot}1, 1{\cdot}2 \ldots$ soon makes it clear that $1{\cdot}1 < x < 1{\cdot}2$.

Using four figure tables we have:

x	$\sin x$	$\frac{1}{x}$		Difference
1·1	0·8912	0·9091	$\sin x < \frac{1}{x}$	0·0179
1·2	0·9320	0·8333	$\sin x > \frac{1}{x}$	0·0987

The differences are in the ratio of 1:6 (nearly). Hence a new estimate of the root is:

$$x \simeq 1{\cdot}1 + \frac{1}{7} \text{ of } 0{\cdot}1 = \text{between } 1{\cdot}11 \text{ and } 1{\cdot}12$$

Proceeding one stage further we have:

x	$\sin x$	$\frac{1}{x}$	Difference
1·11	0·8957	0·9009	0·0052
1·12	0·9001	0·8929	0·0072

$$\therefore x \simeq 1{\cdot}11 + \frac{52}{124} \text{ of } 0{\cdot}01 = 1{\cdot}1142 \text{ (which is correct to 4 places)}$$

3.7 EXAMPLES

1. Sketch the curves:

   (a) The six inverse trigonometrical functions.

   (b) The six inverse hyperbolic functions.

   (c) $y = \log_{\frac{1}{2}} x,\ y = \dfrac{1}{\log x},\ y = x \log x.$

   (d) $y = \dfrac{1}{1 + \cosh x},\ y = \dfrac{1}{1 + \sinh x}$

   (e) $y = \left(1 + \dfrac{1}{x}\right)^x,\ y = \left(1 - \dfrac{1}{x}\right)^x$

   (f) $y = \sin\left(\dfrac{1}{x}\right)$ from $x = \dfrac{1}{4\pi}$ to $x = \infty$

   (g) $y = \tan^{-1}(e^x),\ y = \log \tan\left(\dfrac{x}{2} + \dfrac{\pi}{4}\right)$

   (h) $y = x^x,\ y = \dfrac{1}{x^x},\ y = x^{1/x},\ y = e^{-1/x}$

   (i) $y = +\sqrt{[(x+1)(x+2)]},\ y = \dfrac{1}{\sqrt{(1-x^2)}},\ y = \dfrac{x^3}{(x-1)^{3/2}}$

   (j) $x^3 + y^3 = 1,\ +\sqrt{(x)} + \sqrt{(y)} = +\sqrt{(2)},\ x^2y + xy^2 = 1.$

2. By adding ordinates sketch in succession:

   (a) $1 + x + \dfrac{x^2}{2!};\ 1 + x + \dfrac{x^2}{2!} + \dfrac{x^3}{3!};$ etc. for $-1 < x < +1$

   (b) $x - \dfrac{x^2}{2};\ x - \dfrac{x^2}{2} + \dfrac{x^3}{3};$ etc. for $-1 < x < +3$

   (c) $\sin x - \dfrac{\sin 3x}{3};\ \sin x - \dfrac{\sin 3x}{3} + \dfrac{\sin 5x}{5};$ etc.

   (d) $y = x + \dfrac{20}{x}$ and $x = \dfrac{y}{2} + \dfrac{5}{y}$

3. (a) If $f(x) \equiv$ integral part of x draw $y = f(x)$ and $y = xf(x)$

   (b) If $F(x + 2\pi) \equiv F(x)$ and $F(x) \equiv (x^2/2)$ for $-\pi < x < +\pi$ draw $y = F(x)$ for $-3\pi < x < 5\pi$.

4. Sketch the curves:

   (a) $y = x \log x,\ y = e^{-x} \sin x$

(b) $y^2(4 - x) = x^3$, $y^2 = (x - 3)(x - 1)(x + 1)$
$y^2(2 - x) = x^2(2 + x)$

(c) $y = (1 - \sin x)/(\pi - 2x)$ [See 1.2.2]

5. Find graphically the range of values of x for which:

(a) $\dfrac{x^2 + 1}{x^2 - 1} > 2$

(b) $1 - x^2 - x^3 < 0$

(c) $\sin x > \tanh x$

6. Find graphically the number of real roots (and their rough values) of the equations:

(a) $x^5 - 3x + 2 = 0$

(b) $2x = \tan x$

(c) $10 - x - \log_{10} x = 0$

(d) $2 + \log x - e^x = 0$

(e) $2x \cot x = x^2 - 1$

(f) $y = x + \dfrac{20}{x}$ and $x = \dfrac{y}{2} + \dfrac{5}{y}$ (See Question 2 (d))

7. Use linear interpolation to improve your estimates of the roots of equations in Question 6.

4
Iterative Methods

4.1 THE INDIRECT APPROACH

This chapter deals with problems in which the required solutions are difficult or impossible to calculate directly and introduces methods by which the solution may be obtained to a known degree of accuracy. An essential property of any of these so-called approximate methods is that there should be, in general, no theoretical limit to the degree of accuracy it will give.

Instead of calculating a value x directly the method consists of calculating a convergent sequence of numbers $x_0, x_1, x_2 \ldots x_n \ldots$ such that $x_n \to x$, the required value, as $n \to \infty$. Then for some value of n the difference between x_{n+1} and x_n will be negligible in accordance with the degree of accuracy required by the particular problem, and x_n is taken to be the required answer. It is important to make full use of repetition in calculating the successive approximations x_1, x_2 etc. so that if x_1 is calculated from x_0 by a certain sequence of operations then x_2 is calculated from x_1 by the same sequence of operations. In precisely the same way x_3 is obtained from x_2, and so on.

Plainly it will be an advantage to use a method which provides a rapidly convergent sequence, and the process is further speeded up if a good approximation can conveniently be found as the starting point x_0.

In the next section the problem of finding the square root of a number is used to illustrate this method of indirect approach.

4.2 SQUARE ROOTS

If $xy = N$ then x and y are called factors of N and if $x^2 = N$ then x is called the square root of N. Thus the problem of finding the square root of a number can be regarded as the problem of expressing the number as a product of two *equal* factors, e.g. 12·25 is exactly equal to $3{\cdot}5 \times 3{\cdot}5 \;\therefore\; \sqrt{12{\cdot}25} = 3{\cdot}5$.

This however cannot be done with the majority of numbers, and we often need to be able to calculate square roots to any required number of decimal places, e.g. there is no decimal number which is exactly equal to the square root of 10. We have instead:

$$3{\cdot}1 \times 3{\cdot}1 < 10 < 3{\cdot}2 \times 3{\cdot}2$$
$$3{\cdot}16 \times 3{\cdot}16 < 10 < 3{\cdot}17 \times 3{\cdot}17$$
$$3{\cdot}162 \times 3{\cdot}162 < 10 < 3{\cdot}163 \times 3{\cdot}163$$

i.e. $\sqrt{10}$ is between 3·162 and 3·163.

4.2.1 A quicker and easier way of obtaining a more accurate value of $\sqrt{10}$ can be found by a return to the definition of §4.2 and an attempt to find the solution of the simultaneous equations $xy = 10$ and $x = y$, using a geometrical method. The solution of the equations is of course $x = y = \sqrt{10}$ and is represented by the point R, in Fig. 4.2.1, at which the straight line $y = x$ crosses the hyperbola $xy = 10$. OA $= x_0$ is approximately equal to $\sqrt{10}$. P is on the curve $xy = 10$ $\therefore$ AP $= 10/x_0$. There is another point Q on the curve whose co-ordinates are $x = 10/x_0$ and $y = x_0$ $\therefore$ BQ $= x_0$. There is symmetry

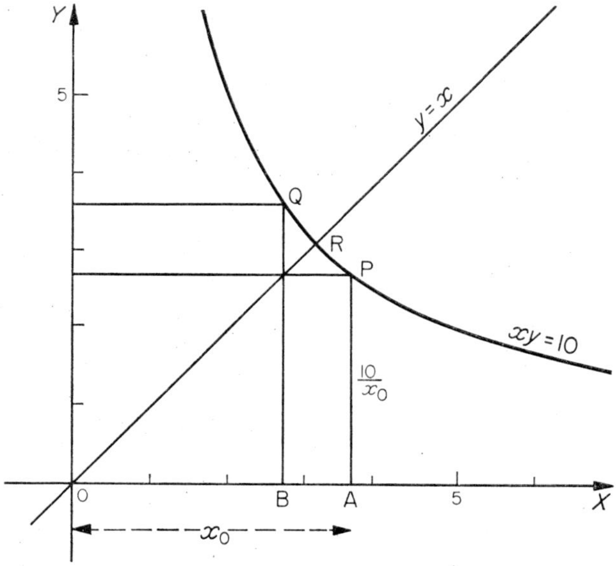

Fig. 4.2.1

about the line $y = x$ and consequently R is the mid-point of the arc PQ. We cannot find the co-ordinates of R, but the mid-point of the straight line joining P and Q is a point M on the straight line $y = x$ and is clearly very much nearer to R than either P or Q. [M is not shown in diagram].

From the trapezium APQB the ordinate of M is equal to $\frac{1}{2}$(AP + BQ) $= \frac{1}{2}[(10/x_0) + x_0]$. $\therefore$ $\frac{1}{2}[(10/x_0) + x_0]$ is not exactly equal to $\sqrt{10}$ but is clearly a much closer approximation than was the original estimate x_0.

If we begin with $x_0 = 3{\cdot}162$ as a first estimate of the value of $\sqrt{10}$ (the error is $< 0{\cdot}001$, see §4.2) then our next estimate will be:

$$x_1 = \tfrac{1}{2}\left(\frac{10}{3{\cdot}162} + 3{\cdot}162\right) = \tfrac{1}{2}(3{\cdot}162\,555\,3 + 3{\cdot}162)$$

$$= 3{\cdot}162\,277\,6\ldots$$

and we can verify by squaring that

$$3{\cdot}162\,277 < \sqrt{10} < 3{\cdot}162\,278.$$

By repeating the same argument we could now obtain further accuracy by calculating the value of $x_2 = \frac{1}{2}[(10/x_1) + x_1]$, where x_1 is put equal to 3·162 278. $\therefore$ $10/x_1$ by division $=$ 3·162 277 320. Thus confirming that x_1 gave $\sqrt{10}$ correctly to 7S and leading to $x_2 = \frac{1}{2}[(10/x_1) + x_1] = 3{\cdot}162\,277\,660\ldots$

4.2.2 Geometrical considerations lead us to see that $\sqrt{10}$ is nearly equal to the average of 3·162 and 10/3·162, and we now analyse the value of these two numbers using an algebraic approach:

Let $\sqrt{10} = 3{\cdot}162 + h$, i.e. $3{\cdot}162 = \sqrt{(10)} - h$ and since we already know that $|h| < 0{\cdot}001$ we have

$$\frac{10}{3{\cdot}162} = \frac{10}{\sqrt{(10)} - h}$$

$$= \frac{\sqrt{10}}{1 - \dfrac{h}{\sqrt{10}}} \text{ [dividing numerator and denominator by } \sqrt{10}]$$

$$= \sqrt{10}\left[1 + \frac{h}{\sqrt{10}} + \frac{h^2}{10} + \ldots + \left(\frac{h}{\sqrt{10}}\right)^n + \ldots\right]$$

using the binomial theorem

$$\therefore \frac{10}{3{\cdot}162} = \sqrt{(10)} + h + \frac{h^2}{\sqrt{10}} + \ldots$$

$\therefore$ the error in saying that $10/3{\cdot}162 \simeq \sqrt{(10)} + h$ is equal to a G.P. whose first term is $h^2/\sqrt{10}$ and whose common ratio is $h/\sqrt{10}$ i.e. the error is $(h^2/\sqrt{10})\,.\,1/[1 - (h/\sqrt{10})] < 0{\cdot}000\,000\,3$ [using $|h| < 0{\cdot}001$] and does not affect the first six decimal places.

We therefore have:

$$\frac{10}{3{\cdot}162} \simeq \sqrt{(10)} + h$$

$$\text{and } 3{\cdot}162 = \sqrt{(10)} - h$$

adding gives

$$\sqrt{10} = \tfrac{1}{2}\left(\frac{10}{3{\cdot}162} + 3{\cdot}162\right)$$

$$= 3{\cdot}162\,278 \text{ to 6 D (as in §4.2.1).}$$

Similarly it can be shown that if x_0 is a sufficiently good approximate value of $\sqrt{N}$ then x_1 is a better one, where $x_1 = \frac{1}{2}[(N/x_0) + x_0]$, and that

$x_2 = \frac{1}{2}[(N/x_1) + x_1]$ is better still. We can use the repetitive formula $x_{r+1} = \frac{1}{2}[(N/x_r) + x_r]$ until the desired degree of accuracy has been reached. The procedure is outlined in the flow chart:

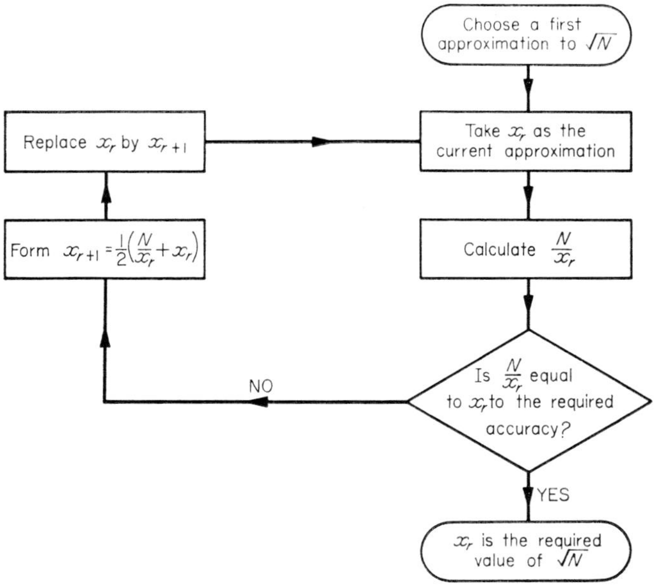

Worked Example:

Calculate $\sqrt{\pi}$ as accurately as possible given that $\pi = 3{\cdot}141\ 592\ 654$ to 10S and four figure tables give $\sqrt{\pi} = 1{\cdot}772$.

r	x_r	π/x_r	$x_{r+1} = \frac{1}{2}\left(\frac{\pi}{x_r} + x_r\right)$
0	1·772	1·772 908	1·772 454
1	1·772 454	1·772 453 702	1·772 453 851
2	1·772 453 851	1·772 453 851	—

$$\therefore \sqrt{\pi} = 1{\cdot}772\ 453\ 851 \text{ (to 10S)}$$

A more analytical discussion of this method of successive approximation to square, and higher roots of a number follows in §4.3.

4.3 GENERAL METHOD

If x is an approximate value of $\sqrt[n]{N}$ and if $x + h$ is equal to $\sqrt[n]{N}$ then h is small compared with x and we have:

$$N = (x + h)^n$$

Using the binomial theorem this gives:

$$N = x^n + nx^{n-1} \,.\, h + \text{terms in } h^2, h^3, \ldots$$

$\therefore\ N - x^n \simeq nx^{n-1} \,.\, h$ if h is small enough to make the terms in h^2, h^3, negligible, and this is valid because in practice we would use quite small values of h.

$\therefore\ h$ is approximately equal to $\dfrac{N - x^n}{nx^{n-1}}$

$\therefore$ a new approximation to $\sqrt[n]{N}$ is:

$$\sqrt[n]{N} = x + h$$

$$= x + \frac{N - x^n}{nx^{n-1}}$$

which can be written as $x + \dfrac{1}{n}\left[\dfrac{N}{x^{n-1}} - x\right]$.

To calculate a series of approximations to $\sqrt[n]{N}$ we can therefore use the repetitive formula

$$x_{r+1} = x_r + \frac{1}{n}\left[\frac{N}{x_r^{\,n-1}} - x_r\right] \qquad (1)$$

with r put equal to 0, 1, 2, . . . in succession.

Worked example

Calculate $\sqrt[3]{1001}$ starting with a first estimate of 10 and using $x_{r+1} = x_r + \frac{1}{3}[(N/x_r^2) - x_r]$.

$$\text{let } x_0 = 10$$

$$\text{then } x_1 = x_0 + \tfrac{1}{3}\left(\frac{1001}{x_0^2} - x_0\right)$$

$$= 10 + \tfrac{1}{3}(10{\cdot}01 - 10)$$

$$= 10{\cdot}003\,333\ldots$$

If we now put $x_1 = 10{\cdot}003\,33$ and calculate x_2 we can verify the number of correct figures we have obtained, and at the same time find a closer approximation still.

$$\text{let } x_1 = 10{\cdot}003\,33$$

$$\text{then } \frac{N}{x_1^2} = \frac{1001}{100{\cdot}066\,611\,1\ldots} = 10{\cdot}003\,336\,67\ldots$$

$$\therefore\ x_2 = x_1 + \tfrac{1}{3}(0{\cdot}000\,006\,67) = 10{\cdot}003\,332\,22\ldots$$

Since this confirms that x_1 gave the value of $\sqrt[3]{1001}$ correct to 5D as stated, the following places calculated in x_2 are also probably correct, i.e. $\sqrt[3]{1001} = 10{\cdot}003\,332\,2\ldots$

4.3.1 Two useful properties of the general method need emphasising.

There is clearly no limit to the number of decimal places obtainable by continuing the sequence of successive approximations.

It is also plain that the method is self correcting and will still lead to the correct result even if a mistake is accidentally made at any stage.

e.g. if x_1 is miscalculated as 10·003 03 we would then have:

$$x_1 = 10{\cdot}003\ 03$$

$$\frac{N}{x_1^2} = \frac{1001}{100{\cdot}060\ 609\ 2} = 10{\cdot}003\ 936\ 69\ldots$$

$$\therefore\ x_2 = x_1 + \tfrac{1}{3}(0{\cdot}000\ 906\ 69)$$

$$= 10{\cdot}003\ 03 + 0{\cdot}000\ 302\ 23$$

$$= 10{\cdot}003\ 333\ 23$$

which agrees to 7D with the result previously obtained when no mistake was made. This is a general property of iterative methods: any small error will be automatically corrected during further applications of the process. If a really gross error is made the sequence may not converge, but it is usually quite plain to see that such an error has occurred. Appropriate steps can then be taken to correct it.

The number of reliable decimal places which may be retained at any stage is discussed in the next section.

4.3.2 Estimation of errors*

At each stage of a calculation by successive approximation there is an error. In formula (1) of §4.3, x_r is one of the approximations to the value of $\sqrt[n]{N}$ and if h_r is the error at this stage then $\sqrt[n]{(N)} - x_r = h_r$ where h_r is small, and is positive if x_r is less than $\sqrt[n]{N}$. The next estimate x_{r+1} is calculated from the formula

$$x_{r+1} = x_r + \frac{1}{n}\left(\frac{N}{x_r^{n-1}} - x_r\right)\ldots \qquad (1)\ (\S 4.3)$$

We can examine this value of x_{r+1} and hence the corresponding error h_{r+1} by analysing the value of the fraction N/x_r^{n-1}

We have $$\frac{N}{x_r^{n-1}} = \frac{N}{[\sqrt[n]{(N)} - h_r]^{n-1}} \text{ (definition of } h_r\text{)}$$

$$= \frac{N}{N^{\frac{n-1}{n}}\left(1 - \dfrac{hr}{\sqrt[n]{N}}\right)^{n-1}}$$

$$= \sqrt[n]{N}\left[1 - \frac{h_r}{\sqrt[n]{N}}\right]^{-(n-1)}$$

* May be omitted on first reading

Using the binomial theorem this gives:

$$\sqrt[n]{N}\left[1+\frac{(n-1)\,h_r}{\sqrt[n]{N}}+\frac{(n-1)\,n\,.\,h_r^2}{1.2\,(\sqrt[n]{N})^2}+\ldots\right]$$

$$=\sqrt[n]{N}+(n-1)\,h_r+\frac{n\,(n-1)}{2}\cdot\frac{h_r^2}{\sqrt[n]{N}}+\ldots$$

$$=x_r+n\,h_r+\frac{n\,(n-1)}{2}\cdot\frac{h_r^2}{\sqrt[n]{N}}+\ldots$$

Substituting this value in formula (1) from §4.3 we have:

$$x_{r+1}=x_r+\frac{1}{n}\left[n\,h_r+\frac{n\,(n-1)}{2}\cdot\frac{h_r^2}{\sqrt[n]{N}}+\ldots\right]$$

$$=\sqrt[n]{N}+\frac{n-1}{2}\cdot\frac{h_r^2}{\sqrt[n]{N}}+\ldots$$

This shows that whatever the sign of h_r may be, x_{r+1} is greater than $\sqrt[n]{N}$ by an amount h_{r+1} which is approximately equal to $(n-1)\,.\,h_r^2/2\sqrt[n]{N}$ provided that the neglected terms in h_r^3 and higher powers are small enough.

This result $h_{r+1}\simeq\{(n-1)/2\,.\,\sqrt[n]{N}\}\,.\,h_r^2$ enables us to determine the number of decimal places we can rely on when calculating x_{r+1} even though the exact values of $\sqrt[n]{N}$ and h_r are still unknown. For this purpose we only need to estimate the size of the error. The difference between x_r and x_{r+1} (i.e. the correction applied at this stage) is nearly equal to h_r and x_r and x_{r+1} are nearly equal to $\sqrt[n]{N}$.

Therefore the error is approximately equal to

$$\frac{n-1}{2}\cdot\frac{(x_r-x_{r+1})^2}{x_r}.$$

Referring back to the calculation of $\sqrt[3]{1001}$ in 4.3 we have

r	x_r	x_{r+1}	Estimate of error	Correct places in x_{r+1}
0	10	10·003 333	$\frac{2}{2}\cdot\frac{(0{\cdot}003\ 333)^2}{10}\simeq 1\times 10^{-6}$	5
1	10·003 33	10·003 332 22	$\frac{2}{2}\cdot\frac{(2{\cdot}2\,.\times 10^{-6})^2}{10}\simeq 5\times 10^{-13}$	11

The estimated errors show that $x_1 = 10{\cdot}003\ 33$ to 5D and that the calculation of x_2 could have been accurately carried out to 11 decimal places.

Since $x_{r+1} > \sqrt[n]{N}$ for all values of $r > 0$ it follows that $x_1\,x_2\,x_3\ldots x_r\ldots$ is a decreasing sequence, and the convergence to the limit $\sqrt[n]{N}$ is rapid because the error at any stage is very nearly proportional to the square of the previous error.

This analysis of errors while of real theoretical interest will not be used a great deal. In practice it is usual, in most cases, to compute an additional stage of the process in order to be certain that the required accuracy has been obtained.

4.3.3 Examples

Calculate (a) $\sqrt[4]{3}$, (b) $\sqrt[3]{25}$, (c) $\sqrt{e}$ as accurately as your equipment will permit and find the number of decimal places which are correct at each stage of your work.

4.4 CONVERGENCE OF ITERATIVE METHODS IN THE SOLUTION OF EQUATIONS

Quadratic, cubic and quartic equations are dealt with fully in algebra text books. The standard methods of solving cubics and quartics although very interesting and instructive tend to be long and in the case of the quartic can be very involved. There is no general algebraic method of solving equations of higher degree than these. In addition, many equations involve trigonometrical, exponential, or other functions and two different kinds of function may occur in the same equation, e.g.

$$x - \sin x = \frac{2\pi}{n} \text{ (Area of segment of circle)}$$

$$\cosh x \cos x = -1 \text{ (Vibration theory)}$$

We need to develop general methods which can be applied to solve any form of equation.

Numerical methods of calculating the roots of these equations depend upon first finding how many real roots there are, and their approximate values (as in Chapter 3). Having found a first approximation to a root of an equation better accuracy can then be obtained as shown in the following sections.

4.4.1 A special iterative method which is of real interest, and is very useful in particular cases is now introduced. (A more general method is discussed later).

Any algebraic equation can be expressed in the form $x = f(x)$, often in several different ways. For example $x^3 - 3x - 20 = 0$ can be written in the forms:

(a) $x = \sqrt[3]{(3x + 20)}$ (b) $x = \frac{1}{3}(x^3 - 20)$

(c) $x = \dfrac{20}{x^2 - 3}$ (d) $x = \sqrt{\left(3 + \dfrac{20}{x}\right)}$

Having expressed an equation in the form $x = f(x)$ the formula $x_{r+1} = f(x_r)$ can be used repeatedly to calculate the sequence $x_0, x_1, x_2, x_3, \ldots x_n, \ldots$

In some cases this sequence converges to a definite limit, in other cases it may diverge and we will examine later the question of testing for convergence. But clearly if the sequence does converge then its limiting value will satisfy the equation $x = f(x)$ and hence be one of the roots of the original equation.

The methods of Chapter 3 show that the equation $x^3 - 3x - 20 = 0$ has only one real root and that this root is near $x = 3$. Choosing the form in (a), $x = \sqrt[3]{(3x + 20)}$, we use the repetitive formula $x_{r+1} = \sqrt[3]{(3x_r + 20)}$.

$$\text{Put } x_0 = 3 \quad \text{then } x_1 = \sqrt[3]{29} = 3{\cdot}072$$

$$\text{Put } x_1 = 3{\cdot}072 \text{ then } x_2 = \sqrt[3]{29{\cdot}216} = 3{\cdot}080$$

$$\text{Put } x_2 = 3{\cdot}08 \quad \text{then } x_3 = \sqrt[3]{29{\cdot}24} = 3{\cdot}081$$

This is as far as four figure tables will take us but the root is clearly very near to $x = 3{\cdot}08$.

Examples

1. Continue the above calculation using six figure tables (or a hand machine) to obtain the root to four decimal places.
2. Try the other forms (b), (c), (d) in which the same equation can be expressed. Which of them give rise to convergent sequences?

4.4.2 The iterative formula $x_{r+1} = f(x_r)$, is useful in solving an equation which has been re-arranged in the form $x = f(x)$.

If α is a root of the equation $x = f(x)$ then $\alpha = f(\alpha)$ exactly. If also x_r is one of the successive approximations obtained by using the formula $x_{r+1} = f(x_r)$ then, if h_r is the error at this stage, we let $x_r = \alpha + h_r$, where h_r is of course small compared with α and x_r. (Note that $x_r > \alpha$ if h_r is positive).

We then have

$$\begin{aligned} x_{r+1} &= f(x_r) \\ &= f(\alpha + h_r) \text{ [Taylor's theorem, see 1.3]} \\ &= f(\alpha) + h_r f'(\alpha) + \frac{h_r^2}{2!} f''(\alpha) + \ldots \\ &\simeq \alpha + h_r f'(\alpha) \end{aligned}$$

provided that the terms in h_r^2 and higher powers of h_r are small enough to be neglected.

i.e.
$$h_{r+1} = x_{r+1} - \alpha \simeq h_r \,.\, f'(\alpha) \ldots \tag{2}$$

$$\therefore |h_{r+1}| < |h_r| \text{ provided } |f'(\alpha)| < 1 \ldots \tag{3}$$

i.e. the error in x_{r+1} is less than the error in x_r whenever $f'(\alpha)$ is numerically less than 1, and that the convergence of the sequence $x_0, x_1, x_2, \ldots$ will be rapid if $|f'(\alpha)|$ is small compared to 1, and will be slow if $|f'(\alpha)|$ is near 1.

It also follows from equation (2) (above) that when $f'(\alpha)$ is positive the sequence will be an increasing or decreasing one, and that when $-1 < f'(\alpha) < 0$ the sequence will oscillate alternately above and below the limit towards which it converges.

In using this test for convergence we must remember that α and hence $f'(\alpha)$ are both unknown when we want to use the test.

Luckily in all straight forward cases the value of $f'(x)$ remains fairly constant for all values of x near α, consequently $f'(x_r)$ is not very different in value from $f'(\alpha)$ and is used instead. For this reason it follows from equation (2) above that $(h_{r+1}/h_r) \simeq f'(x_r)$ which remains nearly constant as r increases, i.e. the errors in the successive approximations form very nearly a Geometric Series with (almost) constant common ratio $\simeq f'(\alpha)$ or $f'(x_r)$.

Example Estimate the value of $f'(\alpha)$ in the equations (a), (b), (c), (d) of §4.4.1 and refer back to your answer to example 2 of §4.4.1.

4.4.3 Worked example: Solve the equation $x^4 + x^2 = 80$.

By inspection, this equation clearly has two roots near $x = \pm 3$ and such that $x^2 < 9$ and $71 < x^4 < 80$. A simple sketch of $y = x^4$ and $y = 80 - x^2$ will show that these are the only two real roots.

Some alternative re-arrangements of the equation are shown below and we will evaluate the root near $x = +3$.

(a) $$x = \sqrt[4]{(80 - x^2)}$$

Here $f(x) = (80 - x^2)^{\frac{1}{4}}$

$$\therefore f'(x) = \tfrac{1}{4}(80 - x^2)^{-\frac{3}{4}} . (-2x) = -\frac{x}{2 . (80 - x^2)^{\frac{3}{4}}}$$

$$\therefore f'(\alpha) \simeq -\frac{3}{2 . (71)^{\frac{3}{4}}} \simeq -0{\cdot}06$$

This way would lead to oscillation and rapid convergence.

(b) $$x = \sqrt{(80 - x^4)}$$

Here $f(x) = \sqrt{(80 - x^4)}$

$$\therefore f'(x) = \tfrac{1}{2}(80 - x^4)^{-\frac{1}{2}} . (-4x^3) = \frac{-2x^3}{\sqrt{(80 - x^4)}}$$

$$\therefore f'(\alpha) \simeq -\frac{2.27}{\sqrt{(9)}} = \text{about } -18$$ [using $80 - x^4 = x^2$ to give $f'(x) = -2x^2$]

This way would therefore be quite useless.

(c) $$x = \sqrt{\left(\frac{80}{1 + x^2}\right)}$$

Here $f(x) = \dfrac{\sqrt{(80)}}{\sqrt{(1 + x^2)}}$

$$\therefore f'(x) = -\frac{\sqrt{(80)}\,.\,(2x)}{2(1+x^2)^{3/2}} = -\frac{\sqrt{(80)}\,.\,x}{(1+x^2)^{3/2}}$$

$$\therefore f'(a) \simeq -\frac{9.3}{10^{3/2}} \simeq -0{\cdot}9$$

This way would lead to oscillation and very slow convergence.

The calculation using arrangement (a) could be carried out as follows:

$$x_{r+1} = \sqrt[4]{(80 - x_r^2)}, \text{ let } x_0 = 3$$

r	x_r	$80 - x_r^2$	$\sqrt{(80 - x_r^2)} = x_{r+1}^2$	x_{r+1}	
0	3·	71	8·426	2·902	Using four figure tables
1	2·9	71·59	8·461	2·909	
2	2·909	71·539	8·458	2·908	
3	2·908	71·543 536	8·458 341	2·908 3	Using a hand calculating machine
4	2·908 3	71·541 791	8·458 238	2·908 305	
5	2·908 305				

The calculation using arrangement (c) could proceed as follows:

$$x_{r+1} = \sqrt{\left(\frac{80}{1 + x_r^2}\right)}$$

r	x_r	$1 + x_r^2$	$80 \div (1 + x_r^2)$	x_{r+1}	
0	3	10	8	2·828	Using four figure tables
1	2·828	9	8·889	2·982	
2	2·982	9·889	8·089	2·844	
3	2·844	9·089	8·802	2·966	
4	2·966				

This sequence is oscillating, and is converging very very slowly as expected, but further use can be made of it as shown in §4.4.4.

It will be noticed that this method shares the general property of all iterative methods that the process is self correcting. Any large mistake in working will be noticed immediately and any small mistake will be automatically removed in the next stage of the calculation. In fact it is frequently an advantage to adjust calculated values of some of the intermediate approximations.

4.4.4 Consider again the previous example which used arrangement (c) $x = \sqrt{\{80/(1 + x^2)\}}$ where we found that the sequence $x_0, x_1, x_2, \ldots$ oscillated while very slowly converging as illustrated in Fig. 4.4.4.

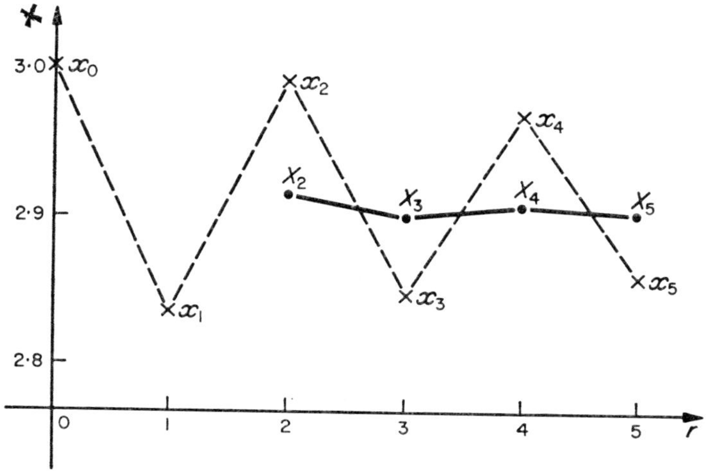

Fig. 4.4.4

$x_0, x_2, x_4 \ldots$ form a slowly decreasing sequence
$x_1, x_3, x_5 \ldots$ form a slowly increasing sequence and the common limit towards which they both eventually approach is plainly nearly midway between each pair x_r and x_{r+1} of consecutive calculated values.

Instead of calculating x_2 we could use $X_2 \simeq \frac{1}{2}(x_0 + x_1)$ as our next approximation, from which we calculate X_3. The next step would then be to use a value for X_4 obtained by averaging X_2 and X_3. The calculation could therefore proceed as follows:

$$x_{r+1} = \sqrt{\left(\frac{80}{1 + x_r^2}\right)} \qquad \text{let } x_0 = 3$$

then $x_1 = \sqrt{(8)}$ $\qquad = 2{\cdot}828 \ldots$ as before

Now use $X_2 = \frac{1}{2}(x_0 + x_1)$ $\qquad = 2{\cdot}91$ (say)

$$\therefore\ X_3 = \sqrt{\left(\frac{80}{9{\cdot}4681}\right)} = \sqrt{8{\cdot}449\,425} \qquad = 2{\cdot}906\,79 \ldots$$

Now use $X_4 = \frac{1}{2}(X_2 + X_3)$ $\qquad = 2{\cdot}908\,4$ (say)

$$\text{and } X_5 = \sqrt{\left(\frac{80}{1 + X_4^2}\right)} = \sqrt{(8{\cdot}457\,740\,92)} \qquad = 2{\cdot}908\,219 \ldots$$

Now use $X_6 = \frac{1}{2}(X_4 + X_5)$ $\qquad = 2{\cdot}908\,31$

which is in fact correct to 5 significant figures.

Examples

Use the method of this section to find, to 4S, solutions of the following equations. In each case a first approximation to the required root is given:

1. $x \sin x = 1$ $(x \simeq 1{\cdot}1)$
2. $1 + \sin^2 x = x^2$ $(x \simeq 1{\cdot}3)$
3. $e^x \tan x = 1$ $(x \simeq 0{\cdot}6)$
4. $1 + x^2 - x^3 = 0$ $(x \simeq 1{\cdot}5)$

4.5 SIMULTANEOUS EQUATIONS

The iterative process for the solution of equations can easily be adapted and used to solve certain simultaneous equations. A simple example will make the method clear enough.

Consider the equations
$$\left.\begin{aligned} x^2 - xy + 20 &= 0 \\ y^2 - 2xy + 10 &= 0 \end{aligned}\right\} \quad \begin{aligned}&(1)\\&(2)\end{aligned}$$

These can be re-arranged as follows:

$$(1)\ y = x + \frac{20}{x} = f(x) \text{ (say), so that } f'(x) = 1 - \frac{20}{x^2}$$

$$(2)\ x = \frac{y}{2} + \frac{5}{y} = F(y) \text{ (say), so that } F'(y) = \tfrac{1}{2} - \frac{5}{y^2}$$

The methods of Chapter 3 can be used to find approximate solutions of these equations, [see §3·7, example no. 6(f)], and one solution will be found to be near $x = 5$, $y = 9$. Using these values we have:

$f'(x) \simeq \frac{1}{5}$ and $F'(y) < \frac{1}{2}$ and both of these indicate good convergence.

We therefore use the iterative formulae:

$$y_r = x_r + \frac{20}{x_r} \text{ and } x_{r+1} = \frac{y_r}{2} + \frac{5}{y_r} \text{ and begin with } x_0 = 5.$$

$$\therefore\ y_0 = 5 + \frac{20}{5} = 9 \text{ hence } x_1 = \frac{9}{2} + \frac{5}{9} = 5{\cdot}06 \text{ approximately}$$

$$\therefore\ y_1 = 5{\cdot}06 + \frac{20}{5{\cdot}06} = 9{\cdot}0126 \text{ hence } x_2 = \frac{9{\cdot}0126}{2} + \frac{5}{9{\cdot}0126} = 5{\cdot}061\,08\ldots$$

let $x_2 = 5{\cdot}0611$

$$\text{then } y_2 = 5{\cdot}0611 + \frac{20}{5{\cdot}0611} = 9{\cdot}012\,81\ldots$$

$$\text{and } x_3 = \frac{9{\cdot}012\,81}{2} + \frac{5}{9{\cdot}012\,81} = 5{\cdot}061\,171\ldots$$

The next stage gives $y_3 = 9{\cdot}012\,826\ldots$ and $x_4 = 5{\cdot}061\,178\ \ldots$ and to 4 decimal places we have the approximate solution:

$$x = 5{\cdot}0612,\ y = 9{\cdot}0128.$$

4.5.1 Examples

1. From equations (1) and (2) above, show that $x^2 = 5[1 + \sqrt{(17)}]$ and $y^2 = 10[4 + \sqrt{(17)}]$ and hence check the accuracy of the solution given.

2. Use an iterative method to solve:

(a) $$\left.\begin{aligned} x^2 - 3xy + 7 &= 0 \\ 2x - y + 2 &= 0 \end{aligned}\right\} \quad (\text{near } x = -2,\ y = -2)$$

(b) $$\left.\begin{aligned} x^3 - y &= 100 \\ y^3 - x &= 150 \end{aligned}\right\} \quad (\text{near } x = 5,\ y = 5)$$

3. Draw a sketch to verify that the equations

$$\left.\begin{aligned} x + \sin y &= 1{\cdot}5 \\ y + \sin x &= 1{\cdot}75 \end{aligned}\right\} \quad \text{have a solution which is near } x = \frac{\pi}{3},\ y = \frac{\pi}{4}.$$

Use the iterative method to find these roots accurately to four D.

4. Given that the following equations have a solution such that $9 < x < 10$ $5 < y < 6$, and $3 < z < 4$, find this set of solutions accurately to 4D.

$$\left.\begin{aligned} x^2 - 2xy + 20 &= 0 \\ y^2 - 3yz + 30 &= 0 \\ z^2 - xz + 20 &= 0 \end{aligned}\right\}$$

4.6 NEWTON'S METHOD FOR f(x) = 0

We come to a more general method of solving an equation in one variable. Geometrically it depends on one simple idea, that the tangent at any point of any curve is a close approximation to the curve for a short distance on each side of the point of contact.

Theory

Given $x = a$ is an approximation to one root of the equation $f(x) = 0$. We require a method of calculating a better approximation.

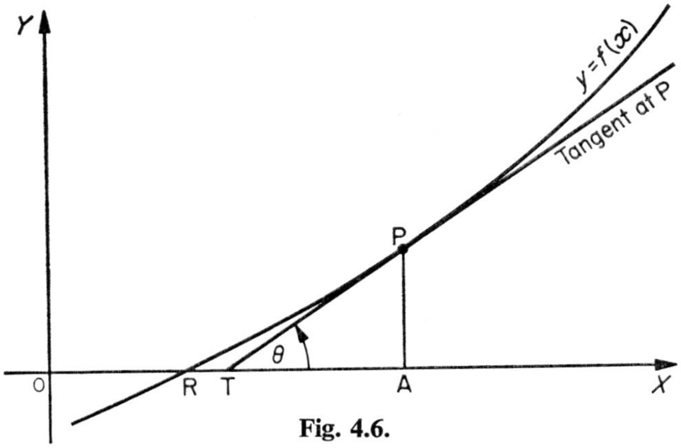

Fig. 4.6.

In the diagram the curve RP represents $y = f(x)$. OR is the exact value of the root of $f(x) = 0$. OA $= a =$ approximate value of the same root. The ordinate at A is AP $= f(a)$ and both RA and AP are small.

The tangent at P is PT whose gradient is $f'(a) = \tan\theta$.

Hence TA $=$ AP $\cot\theta = f(a) \div f'(a)$

$\therefore$ OT $=$ OA $-$ TA $= a - [f(a)/f'(a)]$ and this is in general a closer approximation. Cases in which Newton's Method fails will be discussed later.

Analytically if $x = a$ is an approximation to one root of the equation $f(x) = 0$, then $f(a)$ is small. If the same root is exactly $x = a + h$ then $f(a + h) = 0$ and we need to find the value of h.

Using Taylor's theorem we have:

$$f(a + h) = f(a) + hf'(a) + \frac{1}{2!}h^2 f''(a) + \ldots = 0$$

If the terms involving h^2 and higher powers of h are small enough we have

$$f(a) + hf'(a) \simeq 0$$

i.e. h is approximately equal to $-[f(a)/f'(a)]$ and in general $x = a - [f(a)/f'(a)]$ is a better approximation than $x = a$.

This formula $x = a - [f(a)/f'(a)]$ can be used again with each corrected estimate of the root, so, starting with x_0 instead of a as our first approximation, we have:

$$x_1 = x_0 - \frac{f(x_0)}{f'(x_0)}$$

$$\text{then } x_2 = x_1 - \frac{f(x_1)}{f'(x_1)}$$

$$\text{leading to } x_{r+1} = x_r - \frac{f(x_r)}{f'(x_r)}$$

The following flow diagram and worked examples should make the procedure easy to understand.

In all straight forward cases the successive corrections quickly become smaller and smaller indicating that the sequence $x_0, x_1, x_2 \ldots$ converges rapidly towards an accurate value of the root. Cases where this method fails are easily recognisable because the supposed correction to an approximate root becomes very obviously much too large to be used. These failing cases are dealt with in §4.6.1.

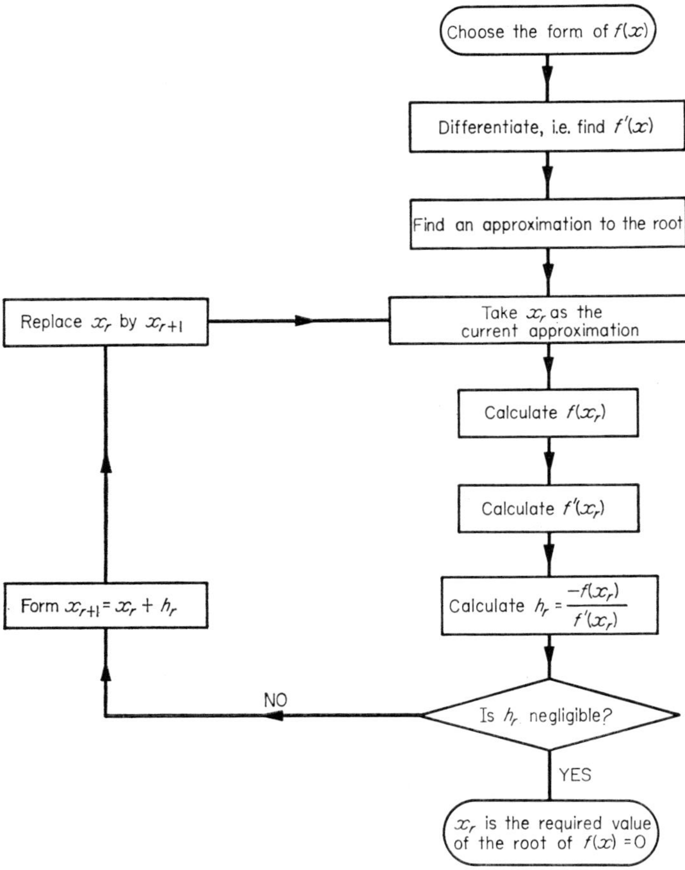

Worked example 1

Find the root of $x^4 + x^2 = 80$ which is near $x = 3$. (See §4.4.3).

let $f(x) \equiv x^4 + x^2 - 80$ then $f'(x) \equiv 4x^3 + 2x$

$= x^2(x^2 + 1) - 80$ $= 2x(2x^2 + 1)$

(a) let $x_0 = 3$ then $f(x_0) = 90 - 80 = 10$

and $f'(x_0) = 6 \times 19 = 114$

$$\therefore x_1 = x_0 - \frac{f(x_0)}{f'(x_0)}$$

$$= 3 - \frac{10}{114}$$

$$= 3 - 0{\cdot}088 = 2{\cdot}912$$

(b) let $x_1 = 2{\cdot}91$

$$\therefore f(x_1) = 80{\cdot}1768 - 80 = 0{\cdot}1768$$

$$\text{and } f'(x_1) = 5{\cdot}82 \times 17{\cdot}9362 = 104{\cdot}388\ldots$$

$$\therefore x_2 = 2{\cdot}91 - \frac{0{\cdot}1768}{104{\cdot}4}$$

$$= 2{\cdot}91 - 0{\cdot}001\,69 = 2{\cdot}908\,31$$

(c) As a check use $x_2 = 2{\cdot}9083$

$$\text{then } f(x_2) = 79{\cdot}999\ 506\ 5 - 90 = -0{\cdot}000\,49\ldots$$

$$\text{and } f'(x_2) = 5{\cdot}8166 \times 17{\cdot}9164\ldots = 104{\cdot}2\ldots$$

$$\therefore x_3 = 2{\cdot}9083 - \frac{-0{\cdot}000\,49}{104}$$

$$= 2{\cdot}9083 + 0{\cdot}000\,005 \text{ (nearly)}$$

which confirms that $x_2 = 2{\cdot}9083$ gave the root correct to 4D and that the error is about 5 in the sixth decimal place.

N.B. Care is necessary in calculating $f(x_1)$, $f(x_2)$ because we require to find the difference between two nearly equal numbers, while only the first 3 or 4 figures matter in $f'(x_1)$, $f'(x_2)$ which remains fairly constant. $f'(x_2)$ could have been taken as 104 in (c) without calculation.

Worked example 2

Evaluate the smallest positive root of the equation $x \sin x = 1$.

(a) A sketch of $y = \sin x$ and $y = 1/x$ easily gives $x \simeq 1{\cdot}1$ (*See* §3·6 *Example* 2).

(b) Using simple linear interpolation as in 3.6.2 we have:

x	$\sin x$	$\frac{1}{x}$		Difference
1·1	0·8912	0·9091	$\sin x < \frac{1}{x}$	0·0179
1·2	0·9320	0·8333	$\sin x > \frac{1}{x}$	0·0987

$$\therefore x \simeq 1{\cdot}1 + \frac{179}{1166} \times 0{\cdot}1$$

$$\simeq 1{\cdot}1 + 0{\cdot}015\ldots = 1{\cdot}115\ldots$$

For a 1st approximation x_0 could be taken as 1·11, or 1·115, or 1·12

(c) Now apply Newton's Method:

$$\left.\begin{aligned} \text{let } f(x) &\equiv \sin x - \frac{1}{x} \\ \text{then } f'(x) &\equiv \cos x + \frac{1}{x^2} \end{aligned}\right\} \text{ and put } x_0 = 1{\cdot}11 \text{ (say)}$$

$$\left.\begin{aligned} f(1{\cdot}11) &= \sin 1{\cdot}11 - \frac{1}{1{\cdot}11} = -0{\cdot}0052 \\ f'(1{\cdot}11) &= \cos 1{\cdot}11 + \frac{1}{(1{\cdot}11)^2} = 1{\cdot}2563 \end{aligned}\right\} \text{using four figure tables.}$$

$$\therefore x_1 = x_0 - \frac{f(x_0)}{f'(x_0)}$$

$$= 1{\cdot}11 - \frac{-0{\cdot}0052}{1{\cdot}26}$$

$$= 1{\cdot}11 + 0{\cdot}0041 = 1{\cdot}114 \text{ (to 4S)}$$

$$\left.\begin{aligned} f(1{\cdot}114) &= 0{\cdot}897\,470 - 0{\cdot}897\,666 = -0{\cdot}000\,196 \\ f'(1{\cdot}114) &= 0{\cdot}441\,075 + 0{\cdot}805\,804 = 1{\cdot}246\,879 \end{aligned}\right\} \text{using six figure tables}$$

$$\therefore x_2 \simeq x_1 - \frac{f(x_1)}{f'(x_1)}$$

$$= 1{\cdot}114 - \frac{-0{\cdot}000\,196}{1{\cdot}247}$$

$$= 1{\cdot}114 + 0{\cdot}000\,157 \simeq 1{\cdot}114\,16 \text{ to 6S}$$

(d) *Check.* $f(1{\cdot}114\,16) = 0{\cdot}897\,540 - 0{\cdot}897\,537 = +\,0{\cdot}000\,003$ this would indicate a further correction of $-\,0{\cdot}000\,003/1{\cdot}25$. i.e. of $-\,2$ or $-\,3$ in the sixth decimal place. No better accuracy is to be expected unless more accurate tables are available. Notice that $f(x_2) = +\,3 \times 10^{-6}$ is calculated as the difference between two numbers which are almost equal, and that we are reduced to only one significant figure at this stage. This one figure is the one which is affected by any rounding-off errors present.

In the above examples full details of the working have been given. In practice it would be better to arrange them in tabular form as follows:

Example 1

r	x_r	$f(x_r)$	$f'(x_r)$	$h_r = \dfrac{-f(x_r)}{f'(x_r)}$	$x_{r+1} = x_r + h_r$
0	3	10	114	−0·088	2·91
1	2·91	0·176	104·4	−0·001 69	2·908 31
2	2·908 3	0·000 49	104·2	<+0·000 005	

Example 2

r	x_r	$f(x_r)$	$f'(x_r)$	$h_r = -\dfrac{f(x_r)}{f'(x_r)}$	$x_{r+1} = x_r + h_r$
0	1·11	−0·005 2	1·256	+0·004 1	1·114 11
1	1·114	−0·000 096	1·2468 . . .	+0·000 157	1·114 157
2	1·114 16	+0·000 003	1·25(nearly)	$\lvert h_2 \rvert$ <0·000 003	

4.6.1 Cases of failure

Newton's formula $x_{r+1} = x_r - [f(x_r)/f'(x_r)]$ will fail whenever the initial approximation x_0 is such that the value of $f(x_0)/f'(x_0)$ is not small enough. A more accurate estimate of x_0 will make the numerator $f(x_0)$ smaller and may therefore be used successfully in some cases. Consider the following variations of the diagram of 4.6.

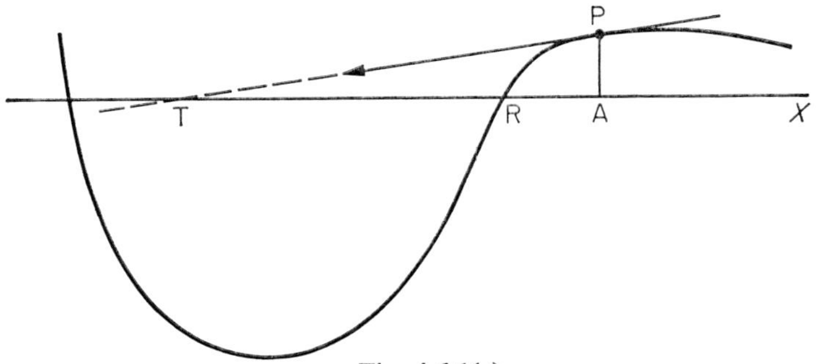

Fig. 4.6.1(a)

In figure (a), the gradient of the tangent at P (where $x = x_0$ the first approximation) is small and it would cross the X axis at T thus making the second estimate of the root worse than the first. In this case a closer approximation than x_0 would have to be used, or better still, an approach should be made to R from the left instead of the right.

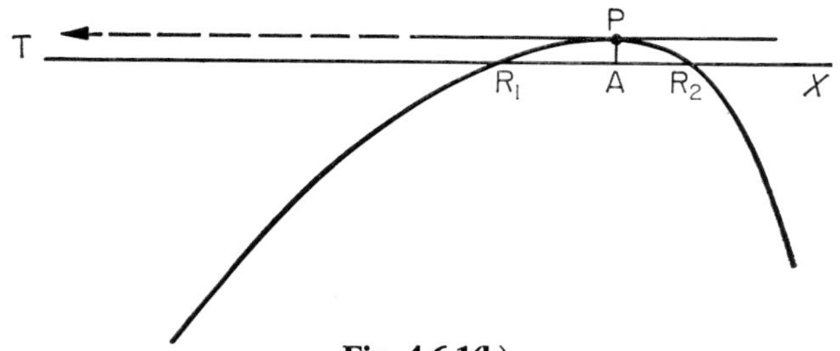

Fig. 4.6.1(b)

Figure (b) illustrates the case where two roots of the equation are close together. Usually the simplest first approximation (A) is nearly midway between the roots (R_1 and R_2) and the tangent at P is very nearly parallel to the X axis. One method of dealing with this situation is now shown:

Worked Example

Consider the equation $x(1 + \sin x) = 1$.

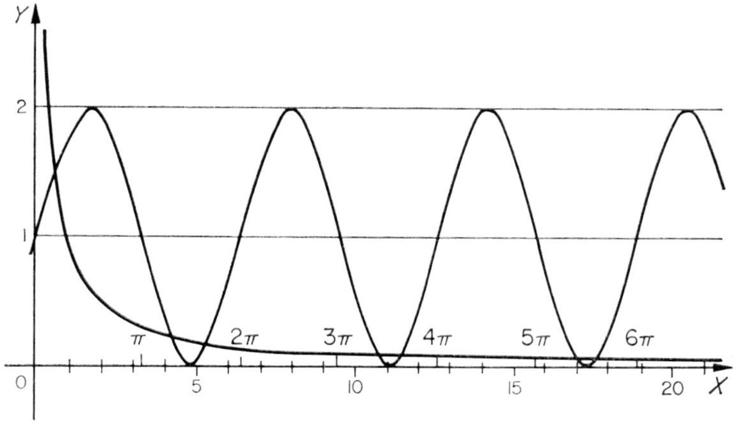

Fig. 4.6.1(c)

(a) The diagram shows $y = 1 + \sin x$ and $y = 1/x$ intersecting near $x = 0{\cdot}5$, 4, 5·5, and with pairs of intersections near $7\pi/2$, and $11\pi/2$, and if n is any integer there will be pairs of roots near $(4n - 1)\pi/2$. Each of these values of x is almost midway between the two roots near it.

(b) We can find a second approximation to these pairs of roots as follows:

$$\text{Let } x = A + h_0 \text{ where } A = \frac{(4n-1)\pi}{2}$$

$$\text{Then the equation } \frac{1}{x} = 1 + \sin x$$

$$\text{becomes } \frac{1}{A+h_0} = 1 + \sin(A + h_0)$$

$$= 1 + \sin(2n\pi - \tfrac{1}{2}\pi + h_0)$$

$$= 1 - \cos h_0$$

$$\simeq \tfrac{1}{2}h_0^2$$

on using the series for cos h_0 and neglecting powers of h_0 higher than the second. Hence since h_0 is small compared with A

$$h_0^2 \simeq \frac{2}{A} \quad \text{i.e.} \quad h_0 \simeq \pm \frac{2}{\sqrt{[(4n-1)\pi]}}$$

e.g. the two roots near $\frac{7\pi}{2}$ are $x = \frac{7\pi}{2} \pm \frac{2}{\sqrt{7\pi}}$

i.e. $x \simeq 10{\cdot}996 \pm 0{\cdot}426$

(c) We can now improve on each of these values of h separately by using Newton's method:

$$\left.\begin{aligned} \text{let } f(h) &\equiv 1 - \cos h - \frac{1}{A+h} \\ \text{then } f'(h) &\equiv \sin h + \frac{1}{(A+h)^2} \end{aligned}\right\} \quad \text{and let } h_0 = +0{\cdot}43$$

$$\therefore f(0{\cdot}43) = 1 - 0{\cdot}9089 - \frac{1}{11{\cdot}43} = +\,0{\cdot}0036$$

$$\text{and } f'(0{\cdot}43) = 0{\cdot}4168 + \frac{1}{11{\cdot}43^2} = 0{\cdot}4245$$

$$\therefore h_1 = 0{\cdot}43 - \frac{0{\cdot}0036}{0{\cdot}424} = 0{\cdot}43 - 0{\cdot}0086 = 0{\cdot}4214$$

To check let $h_1 = 0{\cdot}4214$

then $f(h_1) = 1 - 0{\cdot}9124 - 0{\cdot}0876 = 0$ (except for rounding errors).

Hence the root which is near to and greater than $7\pi/2$ is

$$x \simeq \frac{7\pi}{2} + h$$

$$= 10{\cdot}9956 + 0{\cdot}4214 = 11{\cdot}417$$

4.7 SUMMARY OF SUCCESSIVE APPROXIMATION PROCEDURE

(a) Find a first value graphically.
(b) Find a second estimate by a simple linear interpolation.
(c) Find a sufficiently accurate final value by using an appropriate iterative formula (repeated if necessary).

This procedure may be varied to suit particular examples. Sometimes either (a) or (b) or (c) can be omitted or perhaps the best sequence might prove to be (a), (b), (b) or (a), (c), (c) . . . and so on. The only fixed rule is that once an approximation has been found then, and only then, can a better one be calculated. Time and effort can be saved by a judicious use of stages (a) and, or (b) before eventually using (c).

4.8 EXAMPLES

1. Use the method of §4·3 to establish the formula $x_{r+1} = \frac{1}{2}[x_r + (N/x_r)]$ and prove algebraically that $\frac{1}{2}[x + (N/x)]$ is more nearly equal to $\sqrt{N}$ than is x.

2. Use 6 figure tables to find square roots of 2, 20, 0·3, $\frac{2}{3}$, π, and hence obtain the same square roots to 10S.

3. In the equation $x^3 + 3px + 2q = 0$, if p is positive the real root is given by $x = 2\sqrt{(p)} \sinh \theta$ where $\sinh 3\theta = -[q/(p^{3/2})]$. Calculate the real root of the equation $x^3 + 10x + 3 = 0$:

   (i) using the formula given,
   (ii) using an iterative process.

4. Use an iterative method to solve the equations (a) $xe^x = 1{\cdot}5$, (b) $xe^x = 3$, (c) $xe^x = 12$.

5. Establish the formula $x_{r+1} = x_r(2 - Nx_r)$ as a method of successive approximation to the reciprocal of N, and use it to evaluate $1/\sqrt{2}$ and $1/\pi$ as accurately as your tables will allow.

6. In the example discussed in §4.6.1 re-arrange the equation $x(1 + \sin x) = 1$ in the form $x = \sin^{-1}\{(1 - x)/x\}$, or otherwise, and find a root near $x = 11$ by iteration.

7. How many real roots has the equation $x = \pm\sqrt{(1 - \sin x)}$? Calculate their values as accurately as your tables will permit.

8. Solve the equation $x + 2 = (x - 2)e^{2x}$ using a suitable method of successive approximation.

9. The following equations occur in certain vibration problems:

   (a) $\tan(\frac{1}{2}ml) + \tanh(\frac{1}{2}ml) = 0$. (Show that $\frac{1}{2}ml \simeq 3{\cdot}011.\pi/4$ is the smallest +ve solution)

(b) $\cos ml \cosh ml = -1$. (Show that $ml \simeq 1{\cdot}875$ is the smallest +ve solution).

(c) $7{\cdot}52 \cos mx = \cosh mx$. (Show that $mx \simeq 1{\cdot}305$).

10. (a) Find the range of values of m for which $\sin x = mx$ has more than seven real roots.

(b) find the smallest positive root of the equation $5\pi . \sin x = 2x$.

11. (a) How many roots has the equation $\sinh x = mx$?

(b) Find them all in the case where $m = 10$.

12. Solve the equations $\left.\begin{array}{l} x^3 + y^3 = 1 \\ x^2 + y^2 = 2x \end{array}\right\}$

13. Show graphically that the equation $e^{2x^2} = 16/x$ has a positive root in the range $0 < x < 2$. By the method of inverse interpolation, or otherwise, evaluate this root correct to 7S. (A.E.B.).

14. Sketch the curves $y^2 = x^2 + (1/x)$ and $x^2 = \frac{1}{4}[y^2 + (1/3y)]$ in the positive quadrant and show that they intersect at a point whose x-co-ordinate is approximately unity. Using a double iterative process, evaluate the co-ordinates of this point correct to four decimal places. (A.E.B.).

15. Show graphically that the equation $15x - 10 \sinh x = 1$ has a positive root and evaluate it correct to 6S. (A.E.B.).

16. The equation $x^3 - 3x^2 + 3{\cdot}9 = 0$ has two roots in the neighbourhood of 2.

(a) Evaluate the larger of these correct to 6S, taking all the numerical coefficients as exact.

(b) If 3·9 were a rounded number state briefly how you would investigate the resulting accuracy. (A.E.B.).

17. Show graphically that the equation $\sqrt{(x)} \tanh x + 2x = 5{\cdot}8$ has a real root and evaluate it to 6S. (A.E.B.).

18. Show that a branch of the curve $y = x + (1/x) - 0{\cdot}3$ cuts the circle $x^2 + y^2 = 9$ at two points in the positive quadrant. Find the co-ordinates of the point which has the larger x co-ordinate, correct to 4D. (A.E.B.).

19. Evaluate the two numerically smallest real roots of the equation $x^3 + 2x^2 - 8x = \cosh x$ to 4D.

20. (a) Evaluate $(x \tan x + 1)/(x \sin x + 1)$ where $x = 0{\cdot}463$, to 4D.

(b) Find the cube root of 5765·893 correct to seven significant figures. (A.E.B.).

5
Differences

5.1 DIFFERENCES OF A POLYNOMIAL

Most readers will be familiar with the intelligence test type of problem where the next in a series of numbers must be found.

e.g. 2, 4, 6, 8, . . .

Here each number is found by adding 2 to the number preceding it and so the next one is $8 + 2 = 10$. This change between each number and that preceding it is usually called the difference. In this series the differences are constant, viz. 2, but this is not usually true, for consider the following series of square numbers. (It is convenient to write the numbers of the series on alternate lines and the differences on the line between the corresponding pairs of numbers placed to the right).

	First differences	Second differences
1		
	3	
4		2
	5	
9		2
	7	
16		2
	9	
25		

The differences here are not constant but it will be seen that they themselves increase in steps of 2, i.e. their differences are constant.

We obviously need to distinguish between the sets of differences now being discussed. The differences of the original series of numbers are called first differences, their differences second differences and so on.

This table can be extended by showing the values from which the square numbers are derived. If these first two columns are labelled then it should be clear which columns are first, second and higher differences.

x	x^2		
1	1		
		3	
2	4		2
		5	
3	9		2
		7	
4	16		2
		9	
5	25		

The corresponding table for $4x^2$ for values of x from 0 to 3 is:

x	$4x^2$		
0	0		
		4	
1	4		8
		12	
2	16		8
		20	
3	36		

where again the second differences are constant.

The table for $3x - 5x^2$ for values of x from 0 to 4 is:

x	$3x-5x^2$		
0	0		
		-2	
1	-2		-10
		-12	
2	-14		-10
		-22	
3	-36		-10
		-32	
4	-68		

Note that when finding the difference of two succesive entries in a table as shown above, the upper number is always subtracted from the lower number. This is the reverse of the way in which a subtraction is usually set out. The sign of the difference is important and shows whether the function is increasing or decreasing in the range of values tabulated.

The table for x^3 for values of x from 0 to 4 is:

x	x^3			
0	0			
		1		
1	1		6	
		7		6
2	8		12	
		19		6
3	27		18	
		37		
4	64			

Where it is the third differences which are constant.

5.1.1 Examples

Write down the tables of the following polynomials for $x = 0(1)5$ and difference until a column of constant differences is reached:

1. $2x^3 + x^2 - 3x$
2. $x^2 - 4x - 4$
3. $4x - 3$
4. $2x^4 - x$

5.1.2 Constant Differences of a Polynomial.

It will be seen that there is a connection between the highest power of x that occurs and the order of the differences that are constant. The rule is that *in a polynomial of the* n*th degree the* n*th differences are constant* e.g. the third differences of a cubic and fourth differences of a quartic are always constant.

Note

(i) This rule is only true if the values in a table of the polynomial are exact, that is, the values have not been rounded off. This is discussed later in 5.1.8.

(ii) The rule does not apply to functions which are not polynomials, e.g. consider the table for 2^x for values of x from 0 to 4.

x	2^x				
0	1				
		1			
1	2		1		
		2		1	
2	4		2		1
		4		2	
3	8		4		
		8			
4	16				

Here all columns of differences are exactly the same. (The reader should compare differencing with differentiating. If a polynomial of order n is differentiated n times the result is a constant but if 2^x is repeatedly differentiated the result is always 2^x multiplied by a constant. See §5.5.1.)

Other functions such as sin x, cosh x, log x do not yield a column of constant differences either but only polynomials which are tabulated exactly have this property.

(iii) This rule is also only true of polynomials when they are tabulated at *equal* intervals of the variable. For example:

x	x^3				
0	0				
		1			
1	1		25		
		26		−14	
3	27		11		118
		37		104	
4	64		115		
		152			
6	216				

Here the intervals of the variable are not equal and so the third differences are not constant. In fact no column of differences is constant.

5.1.3 Non-integral values of the variable

So far only integral values of the variable have been used yielding integral differences. If however non-integral values of the variable are used then the differences are also non-integral, e.g. The table for $2x^3$ for $x = 0(0{\cdot}1)0{\cdot}5$.

x	$2x^3$			
0	0			
		0·002		
0·1	0·002		0·012	
		0·014		0·012
0·2	0·016		0·024	
		0·038		0·012
0·3	0·054		0·036	
		0·074		0·012
0·4	0·128		0·048	
		0·122		
0·5	0·250			

Since the function is tabulated to three decimal places each difference is also to three decimal places. Because of this the decimal point may be omitted in the columns of differences and they may then be written as integers. This makes the table easier to write down, easier to read and much neater.

Thus if a function is tabulated to six decimal places the difference written as 2642 has the value 0·002 642. If it is tabulated to three decimal places the difference written as 2642 has the value 2·642.

The above table is then written down in the following form:

x	$2x^3$			
0	0			
		2		
0·1	0·002		12	
		14		12
0·2	0·016		24	
		38		12
0·3	0·054		36	
		74		12
0·4	0·128		48	
		122		
0·5	0·250			

5.1.4 Checking differences

Consider two columns of differences:

a	
	b—a
b	
	c—b
c	
	d—c
d	
	e—d
e	

If the differences in the second column are added together the result is equal to e—a, that is, the difference between the first and last entries in the previous column. Thus the sum of a column of differences, plus the first entry in the previous column, is equal to the last entry in the previous column. This fact should always be used to check that a column of differences has been found correctly. The sums of the first, second and third differences in the table for $2x^3$ are 250, 120 and 36 respectively, which values are seen to check.

5.1.5 Using a machine to calculate differences

It is not necessary to set any number twice if the following method is used:

If the numbers in the column being differenced are positive and increasing or negative and decreasing set the first number in the S.R. and subtract. Set the second number in the S.R. and add. Note down the required difference from the accumulator and clear the accumulator but NOT the S.R. Subtract the second number, which is still in the S.R., set the third number in the S.R. and add etc. The differences in the first case will be positive and in the second negative.

If the numbers are positive and decreasing or negative and increasing add the first number and subtract the second etc. The differences in the first case will be negative and in the second positive.

In some tables there may be either a change in sign of the function values (near a zero value of the function) or previously increasing values will begin to decrease and vice-versa (near a maximum or minimum value). These changes can also occur in any column of differences. When such changes do occur in a column they can be allowed for so that no extra operations are necessary in calculating the differences, but care must be taken to ensure that this is correctly carried out on the machine.

For example in the following table the function is positive and decreasing in the range 0·6 to 1·0 and positive and increasing in the range 1·4 to 1·8. There is as a consequence a change of sign in the first differences.

x	$f(x)$		
0·6	2·656		
		−384	
0·8	2·272		112
		−272	
1·0	2·000		160
		−112	
1·2	1·888		208
		+ 96	
1·4	1·984		256
		352	
1·6	2·336		304
		656	
1·8	2·992		

Having calculated the first three first differences 1·888 will still be set in the setting register and the accumulator cleared. Instead of adding 1·888 into the accumulator as for previous values it is subtracted and 1·984 set and added to give the next difference in the accumulator. Successive differences are calculated in the same way. A similar process will need to be used in calculating second differences when the change in sign in the first differences is reached.

Note that when there is a change in sign in a column the sign of the first positive number should be shown as on the difference 96 above in the first differences.

5.1.6 Checking the tabulation of a function

When a polynomial has been tabulated exactly the fact that there must be a column of constant differences provides a positive check on the tabulation.

Differencing can also be used to check the tabulation of a polynomial which is not exact, or of any function which varies smoothly, since a mistake in a value of the function will cause greater errors in the differences. (See §5.3.3) Such errors will often be easily noted, but not always, so that this method of checking the tabulation, particularly for non-polynomial functions, should be used with caution. Other methods of checking the tabulation of a function are given in chapter 2.

5.1.7 Examples

Tabulate the following polynomials by nesting and difference, writing the differences as integers and checking each column of differences.

1. $x^2 - 2x + 3$ $\qquad x = 0(0{\cdot}2)1{\cdot}0$ to 2D
2. $x^3 + 2x^2 - 4$ $\qquad x = 0(0{\cdot}1)0{\cdot}5$ to 3D
3. $3x^2 - 4x + 5$ $\qquad x = 0(0{\cdot}5)2{\cdot}5$ to 2D
4. $0{\cdot}5x^3 + 1{\cdot}6x^2 - 0{\cdot}7x$ $\qquad x = 0(0{\cdot}1)0{\cdot}5$ to 4D
5. $2{\cdot}3x^3 - 3{\cdot}4x + 4{\cdot}5$ $\qquad x = 0(0{\cdot}2)1{\cdot}0$ to 4D

5.1.8 Rounding off errors

Compare the following tables of the function $f(x) = x^3 + 2x^2 - 1$ and the differences shown, the first table being exact, the second rounded off to 2D.

Table 1

x	$f(x)$			
0	−1·000			
		21		
0·1	−0·979		46	
		67		6
0·2	−0·912		52	
		119		6
0·3	−0·793		58	
		177		6
0·4	−0·616		64	
		241		6
0·5	−0·375		70	
		311		6
0·6	−0·064		76	
		387		6
0·7	+0·323		82	
		469		6
0·8	0·792		88	
		557		6
0·9	1·349		94	
		651		
1·0	2·000			

Table 2

x	$f(x)$							
0	−1·00							
		2						
0·1	−0·98		5					
		7		0				
0·2	−0·91		5		0			
		12		0		2		
0·3	−0·79		5		2		−5	
		17		2		−3		6
0·4	−0·62		7		−1		1	
		24		1		−2		9
0·5	−0·38		8		−3		10	
		32		−2		8		−26
0·6	−0·06		6		5		−16	
		38		3		−8		27
0·7	+0·32		9		−3		11	
		47		0		3		
0·8	0·79		9		0			
		56		0				
0·9	1·35		9					
		65						
1·0	2·00							

It will be seen in table 2, in which the values of the function have been rounded off, that the third differences are no longer constant and that the fourth and higher differences are not zero. The values of the fourth and higher differences arise entirely from rounding off errors and are thus of no use in any calculation. Note how their magnitude increases with the order of difference (considered further below) and that their signs are random. The magnitude of the third differences is smaller than that of the first and second differences and of the fourth and higher differences.

The useful limit of a table

The column of differences which is smallest in magnitude is the highest which is of any use in a calculation and is called *the useful limit* of the table. Usually the average of this column will be near to zero.

Tables of non-polynomials in which the values have been rounded off also have a useful limit which is the column of differences of smallest magnitude. For example third differences are the useful limit of the following table of cot x, since they are of smallest magnitude. Thus fourth and higher differences should be neglected in a calculation using this table.

x	cot x						
0·150	6·616 59						
		4448					
0·151	6·572 11		57				
		4391		1			
0·152	6·528 20		58		−4		
		4333		−3		7	
0·153	6·484 87		55		3		−11
		4278		0		−4	
0·154	6·442 09		55		−1		7
		4223		−1		3	
0·155	6·399 86		54		2		−8
		4169		1		−5	
0·156	6·358 17		53		−3		10
		4116		−2		5	
0·157	6·317 01		51		2		− 8
		4065		0		−3	
0·158	6·276 36		51		−1		
		4014		−1			
0·159	6·236 22		50				
		3964					
0·160	6·196 58						

Maximum errors

Rounding off errors have the greatest effect when entries in the table are in error alternately by $+\frac{1}{2}$ and $-\frac{1}{2}$ in the last decimal place. This is shown in the following table:

Errors in $f(x)$	First Differences	Second	Third	Fourth
$\frac{1}{2}$				
	−1			
$-\frac{1}{2}$		2		
	1		−4	
$\frac{1}{2}$		−2		8
	−1		4	
$-\frac{1}{2}$		2		−8
	1		−4	
$\frac{1}{2}$		−2		
	−1			
$-\frac{1}{2}$				

That is, if the entries in the table have a rounding off error of $\pm\frac{1}{2}$ in the last place then the maximum error in the nth differences is $\pm 2^{n-1}$.

These maximum errors are in practice seldom attained and statistical considerations suggest lower limits which are given in Table B.2 page 55 of Interpolation and Allied Tables. For example, although the maximum error possible in fifth differences due to rounding off errors is 16, the error will only be greater than 11 in about 1% of rounded off fifth differences.

The occurrence of these errors must not be forgotten when using a table which has been rounded off.

5.2 EXTENDING A TABLE OF A POLYNOMIAL

If the degree and the first few differences of a polynomial which has been tabulated exactly are known, then a table of values can be built up by working backwards through the differences, starting with the column of constant differences.

For example, suppose that the following part of the exact table of a cubic is given:

x	$f(x)$			
0	0			
		12		
0·2	0·012		72	
		84		72
0·4	0·096		144	
		228		
0·6	0·324			

Since this is an exactly tabulated cubic it is known that the third differences are constant and thus here they are all equal to 72. The next entries in the other columns are then:

Second differences $144 + 72 = 216$

First differences $228 + 216 = 444$

and so $f(0{\cdot}8) = 0{\cdot}324 + 0{\cdot}444$

$= 0{\cdot}768$

The table can also be extended backwards to give the value of $f(-0{\cdot}2)$ etc. The previous entries in the columns of differences are:

Second differences $72 - 72 = 0$

First differences $12 - 0 = 12$

and so $f(-0{\cdot}2) = 0 - 0{\cdot}012$

$= -0{\cdot}012$

This function tabulated for $x = -\,0{\cdot}4(0{\cdot}2)1{\cdot}0$ is as follows:

x	$f(x)$			
−0·4	−0·096			
		84		
−0·2	−0·012		−72	
		12		72
0	0		0	
		12		72
0·2	0·012		72	
		84		72
0·4	0·096		144	
		228		72
0·6	0·324		216	
		444		72
0·8	0·768		288	
		732		
1·0	1·500			

It is clear from the extension of this table to negative values of x that the magnitudes of the function values are symmetrical about $x = 0$. Further values of the function for negative values of x can be written down from those for the corresponding positive values of x.

Note that if the table is required to be rounded off this must be done *after* it has been completed.

This method of extending a table of a polynomial provides an alternative to that of nesting. If only a few values of the polynomial are required then nesting will be used. For a table of values however it may be easier to find the necessary starting values by nesting and to complete it using differences, checking at some easy value of the variable such as 1·0 or 0·5. If only some values of the polynomial are given, but not its equation, then this method must be used.

5.2.1 Examples

Extend the tables of the following polynomials.

1.

x	0·01	0·02	0·03	0·04
$f(x)$	1·521	1·544	1·569	1·596

Quadratic. Extend to $x = 0{\cdot}10$

2.

x	0	0·2	0·4	0·6	0·8
$f(x)$	4·000	5·168	6·704	8·656	11·072

Cubic. Extend from $-\,1{\cdot}0$ to $2{\cdot}0$

3.

x	0	0·1	0·2	0·3	0·4	0·5
$f(x)$	0	0·001	0·032	0·243	1·024	3·125

Quintic. Extend to $x = 1{\cdot}0$

Tabulate the following polynomials exactly for $x = 0(0{\cdot}1)1{\cdot}0$

4. $8{\cdot}4x^3 - 0{\cdot}2x^2 + 2{\cdot}9x - 0{\cdot}7$
5. $7x^3 - 9x^2 - 7x + 3$

5.3 LOCATING, EVALUATING AND CORRECTING A MISTAKE IN A TABLE

5.3.1 A mistake in calculating or copying down the values of a polynomial is made immediately obvious from the differences. This is equally true for a table which has been rounded off as it is for an exact table, for the magnitude of the error in the differences due to such a mistake is nearly always much greater than rounding off errors.

For example compare the extended table in §5.2 with that below for the same function where $f(0{\cdot}4)$ has been written as 0·086 instead of 0·096:

x	$f(x)$				
—0·4	—0·096				
		84			
—0·2	—0·012		−72		
		12		72	
0	0		0		−10
		12		62	
0·2	0·012		62		40
		74		102	
0·4	0·086		164		−60
		238		42	
0·6	0·324		206		40
		444		82	
0·8	0·768		288		−10
		732		72	
1·0	1·500		360		
		1092			
1·2	2·592				

Note how the effect of the mistake in the table fans out into the differences.

The way in which it builds up is seen in the following table where all entries are zero apart from one which has a mistake m in it.

It will be seen that the mistake builds up in the columns of differences in such a way that the coefficients of m in the columns of differences are the coefficients in the expansion of $(1 - x)^n$ where n is the order of the differences. For example the coefficients of m in the third differences are 1, -3, 3, -1 and $(1 - x)^3 = 1 - 3x + 3x^2 - x^3$. In using this fact to determine a mistake in an exactly tabulated polynomial it is usually easiest to continue differencing until the column of differences which should be all zeros is reached. The pattern of the coefficients corresponding to the order of differences can then be readily picked out and the mistake determined exactly.

	First	Second	Third	Fourth
0				
	0			
0		0		
	0		0	
0		0		m
	0		m	
0		m		$-4m$.
	m		$-3m$	
m		$-2m$		$6m$
	$-m$		$3m$	
0		m		$-4m$
	0		$-m$	
0		0		m
	0		0	
0		0		
	0			
0				

The incorrect value in the table is often indicated by the way in which the incorrect differences fan out as above. If this 'fan' cannot be picked out easily then the incorrect value can be located by noting that for even order differences the largest error is on the same line as the incorrect value in the table, and for odd order differences the largest pair of errors lie on either side of this line.

Having located the mistake the true function value is found by subtracting the value of the mistake, m, from the incorrect value.

If the table does not contain exact values the mistake cannot always be determined exactly, but a good estimate can usually be made. Although there will be no column of constant or of zero differences the errors due to a mistake are usually larger than the rounding off errors (unless the mistake is very small and rounding off errors are near the maximum possible) so that the pattern of coefficients can usually still be seen.

Worked example 1

See the first table in §5.3.1.

Since the function is a cubic and is tabulated exactly the fourth differences should be zero. In fact they are $-10, 40, -60, 40, 10$. The corresponding coefficients for fourth differences are the coefficients of $(1-x)^4$, that is $1, -4, 6, -4, 1$. Comparing these the mistake is seen to have the value -10, i.e. $-0{\cdot}010$ since the table is given to three decimal places.

The incorrect value in this table, 0·086, can be located by the way in which the mistakes fan out, but note that it is on the same line as — 60, the largest incorrect fourth difference.

The true value of the function is now found by subtracting the mistake from the incorrect value i.e. Here the true value is 0·086 — (— 0·010) = 0·096.

This result can be checked by rewriting the table and differencing.

Worked example 2

Find the mistake in the following quartic which is tabulated exactly.

x	$f(x)$					
0	0					
		12				
0·1	0·012		48			
		60		72		
0·2	0·072		120		48	
		180		120		0
0·3	0·252		240		48	
		420		168		0
0·4	0·672		408		48	
		828		216		630
0·5	1·500		624		678	
		1 452		894		−3150
0·6	2·952		1518		−2472	
		2 970		−1578		6300
0·7	5·922		−60		3828	
		2 910		2250		−6300
0·8	8·832		2190		−2472	
		5 100		−222		3150
0·9	13·932		1968		678	
		7 068		456		−630
1·0	21·000		2424		48	
		9 492		504		
1·1	30·492		2928			
		12 420				
1·2	42·912					

Since this function is a quartic and is tabulated exactly the fifth differences should all be zero. The mistake will appear in fifth differences with coefficients those of $(1 - x)^5$, that is, 1, — 5, 10, — 10, 5, — 1. The incorrect fifth differences are 630, — 3150, 6300, — 6300, 3150, — 630 and comparing these with the coefficients the mistake is seen to have the value 630, i.e. 0·630 since the function is tabulated to three decimal places.

The incorrect value in the table, 5·922, can again be located by the way in which the incorrect differences fan out or by the fact that it is on the line between 6300 and − 6300, the largest pair of incorrect fifth differences.

The true value of the function is now found by subtracting 0·630 from 5·922, i.e. 5·292.

This mistake of reversing the order of a pair of digits in a number is one commonly made in copying down a number from a calculating machine or set of tables.

5.3.2 Examples

Find the mistake in each of the following tables:

1.

x	0·4	0·5	0·6	0·7	0·8	0·9	1·0
$f(x)$	5·040	6·375	8·000	9·945	12·420	14·915	18·000
	1·1	1·2	1·3				
	21·525	25·520	30·015		cubic		

2.

x	0	0·1	0·2	0·3	0·4	0·5	0·6
$f(x)$	0	0·002	0·032	0·162	0·512	1·150	2·592
	0·7	0·8	0·9	1·0	1·1		
	4·802	8·192	13·122	20·000	29·282		quartic

3.

x	0	0·2	0·4	0·6	0·8	1·0	1·2
$f(x)$	7·000	7·528	8·344	9·496	11·032	13·000	15·488
	1·4	1·6	1·8	2·0			
	18·424	21·976	26·152	31·000		cubic	

4.

x	0	0·2	0·4	0·6	0·8	1·0	1·2
$f(x)$	2·0000	1·9664	1·8592	1·7888	1·8656	2·2000	2·9024
	1·4	1·6	1·8	2·0	2·2	2·4	2·6
	4·0832	6·4416	8·3216	11·6000	15·7984	21·0272	27·3968
							cubic

5. Quadratic with two mistakes

x	− 1·0	− 0·8	− 0·6	− 0·4	− 0·2	0
$f(x)$	2·4300	0·8764	− 0·4644	− 1·5942	− 2·5076	− 3·2100
	0·2	0·4	0·6	0·8	1·0	
	− 3·6996	− 3·9804	− 4·0404	− 3·8916	− 3.5300	

5.3.3 Mistakes in tables which have been rounded off.

This method can also be applied to tables of polynomials which are not exact or to tables of functions which are not polynomials but in such cases only an estimate of the value of the mistake can be made. If possible, this method should only be used in these cases to locate the incorrect value, and some other method used to correct it.

Worked example 3

Locate the incorrect entry in the following table and estimate its correct value.

x	$f(x)$						
0	−9·800						
		739					
0·1	−9·061		−19				
		720		46			
0·2	−8·341		27		13		
		747		59		−6	
0·3	−7·594		86		7		42
		833		66		36	
0·4	−6·761		152		43		−110
		985		109		−74	
0·5	−5·776		261		−31		149
		1246		78		75	
0·6	−4·530		339		44		−113
		1585		122		−38	
0·7	−2·945		461		6		44
		2046		128		6	
0·8	−0·899		589		12		
		2635		140			
0·9	+1·736		729				
		3364					
1·0	5·100						

The table is differenced until the differences are clearly beginning to increase in magnitude due to the combined effects of rounding off errors and the mistake. There is no column of differences which should be all zeros but for most functions it will be possible to select the column which is smallest in magnitude and which has an average near to zero. (See §5.1.8). The mistake will not affect this average since the sum of the coefficients of $(1 - x)^n$ is always zero; for example $1 - 4 + 6 - 4 + 1 = 0$.

In the above table fourth and fifth differences are the smallest in magnitude and have averages of 13·5 and − 0·2 respectively, so that fifth differences are investigated.

The coefficients of $(1 - x)^5$ which multiply the mistake in the fifth differences are 1, − 5, 10, − 10, 5, − 1 and a similar pattern occurs in the fifth differences, − 6, 36, − 74, 75, − 38, 6. Mistakes of − 6, − 7 and − 8 would give the following errors in the fifth differences:

A mistake of − 6 gives − 6, 30, − 60, 60, − 30, 6.

A mistake of − 7 gives − 7, 35, − 70, 70, − 35, 7.

A mistake of − 8 gives − 8, 40, − 80, 80, − 40, 8.

F

To decide which of these is the most likely value of the mistake find the sum of the "deviations" for each and choose that for which this sum is least. The deviations are the differences in magnitude between each estimated fifth difference and the actual fifth difference. Thus for a mistake of -6 the deviations are $6-6=0$, $36-30=6$, $74-60=14$, $75-60=15$, $38-30=8$, $6-6=0$ which have a sum of 43. Similarly the sums of the deviations for -7 and -8 are 15 and 21 respectively. A mistake of -7 thus gives fifth differences which agree most closely to their actual values. That is, since the table is given to three decimal places, the mistake is estimated to be $-0{\cdot}007$.

The incorrect entry in the table is that on the line between the differences -74 and 75 and is thus $-5{\cdot}776$. It follows that the true value is approximately $-5{\cdot}776-(-0{\cdot}007)=-5{\cdot}769$.

If the function is known analytically it would be preferable to recalculate it at $x=0{\cdot}5$ so that the correction can be made with certainty rather than just estimated. In such cases the main use of this method is in locating the incorrect tabular value.

Worked example 4

Locate the incorrect entry in the following table and estimate its correct value.

x	$f(x)$				
2·698	7·458 67				
		1480			
2·700	7·473 47		3		
		1483		−1	
2·702	7·488 30		2		4
		1485		3	
2·704	7·503 15		5		740
		1490		743	
2·706	7·518 05		748		−2981
		2238		−2238	
2·708	7·540 43		−1490		4477
		748		2239	
2·710	7·547 91		749		−2984
		1497		−745	
2·712	7·562 88		4		744
		1501		−1	
2·714	7·577 89		3		1
		1504		0	
2·716	7·592 93		3		
		1507			
2·718	7·608 00				

The incorrect differences show up very clearly in this table and indicate that he mistake is in f(2·708), which is 7·540 43.

Third differences are the smallest in magnitude and so are investigated. The coefficients for third differences are 1, − 3, 3, − 1 and the corresponding differences are 743, − 2238, 2239, − 745. Mistakes of 745, 746 and 747 would give errors in the third differences thus:

A mistake of 745 gives 745, − 2235, 2235, − 745.

A mistake of 746 gives 746, − 2238, 2238, − 746.

A mistake of 747 gives 747, − 2241, 2241, − 741.

The sums of the deviations, 9, 5 and 13 respectively, indicate that 746 is the best value to take for the mistake. That is, since the table is tabulated to five decimal places, the mistake is estimated to be 0·007 46 giving the correct value of $f(2{\cdot}708)$ to be near 7·540 43 − 0·007 46 = 7·532 97.

This example is an extract from standard tables of cosh x and the mistake is due to copying down the value of the function for $f(2{\cdot}709)$ instead of for $f(2{\cdot}708)$. The correct value, 7·532 96, would in practice be found from the tables.

5.3.4 Examples

Locate and correct the mistake in each of the following tables:

1.

x	0	0·02	0·04	0·06	0·08
$f(x)$	9·900 000	10·028 974	10·164 205	10·306 161	10·455 308
	0·10	0·12	0·14	0·16	0·18
	10·612 110	10·777 035	10·950 558	11·133 116	11·325 205
	0·20	0·22			
	11·527 280	11·739 808		cubic	

2.

x	1·0	1·1	1·2	1·3	1·4	1·5	1·6
$f(x)$	28·800	34·953	42·406	51·336	61·930	74·338	88·917
	1·7	1·8	1·9	2·0			
	105·737	125·079	147·183	172·300		quartic	

3.

x	2·00	2·01	2·02	2·03	2·04	2·05	2·06
$f(x)$	53·000	53·733	54·474	55·221	55·975	56·736	57·509
	2·07	2·08	2·09	2·10			
	58·279	59·061	59·850	60·646		cubic	

4.

x	0·980	0·981	0·982	0·983	0·984	0·985	0·986
$f(x)$	3·3541	3·3793	3·4059	3·4340	3·4638	3·4959	3·5295
	0·987	0·988	0·989	0·990			
	3·5661	3·6056	3·6485	3·6956			

5.

x	1·60	1·61	1·62	1·63	1·64	1·65
$f(x)$	0·8631	0·8506	0·8383	0·8261	0·8142	0·8025
	1·66	1·67	1·68	1·69	1·70	
	0·7906	0·7796	0·7684	0·7574	0·7465	

5.3.5 Two mistakes in a table

If there are two or more mistakes in a table each will affect the differences and, provided that the mistakes are not widely separated, there will be some differences in which errors are caused by both mistakes. It is impossible however for the mistakes to completely overlap in the differences so that it is possible to separate them.

Worked example 5

The following table of exact values of a quadratic contains two mistakes.

x	$f(x)$					
0	8·00					
		−27				
0·1	7·73		6			
		−21		0		
0·2	7·52		6			
		−15		−3	a	
0·3	7·37		3			
		−12		11	$-3a$	b
0·4	7·25		14			
		2		−15	$3a$	$-3b$
0·5	7·27		−1			
		1		9	$-a$	$3b$
0·6	7·28		8			
		9		−2		$-b$
0·7	7·37		6			
		15		0		
0·8	7·52		6			
		21		0		
0·9	7·73		6			
		27				
1·0	8·00					

As before the table is differenced as far as the column of differences which should be all zeros, that is here as far as third differences. The two adjacent entries $f(0{\cdot}4)$ and $f(0{\cdot}5)$ are seen to be in error by the way in which the incorrect differences fan out.

Let the mistakes in $f(0{\cdot}4)$ and $f(0{\cdot}5)$ be a and b respectively. The errors due to a are written in a column with the appropriate coefficients, beginning alongside the first incorrect third difference. Those due to b are entered similarly but beginning alongside the last incorrect difference and working upwards, as shown in the above table. Thus the combined effect of these errors is to produce errors in the third differences of a, $-3a+b$, $3a-3b$, $-a+3b$ and $-b$ which on comparison with the corresponding third differences -3, 11, -15, 9 and

-2 yield $a = -3$ and $b = 2$. The mistakes in $f(0{\cdot}4)$ and $f(0{\cdot}5)$ are thus $-0{\cdot}03$ and $0{\cdot}02$ since the table is given to two decimal places, and the correct values are $f(0{\cdot}4) = 7{\cdot}25 - (-0{\cdot}03) = 7{\cdot}28$ and $f(0{\cdot}5) = 7{\cdot}27 - 0{\cdot}02 = 7{\cdot}25$. Since the values in the table are exact the correction has been made with certainty.

Worked example 6

Locate and correct the two mistakes in the following table of exact values of a cubic.

x	$f(x)$						
−1·0	−1·000						
		−1 875					
−0·5	−2·875		750				
		−1 125		3750			
0	−4·000		4 500		0		
		+3 375		3750			
0·5	−0·625		8 250		−500	a	
		11 625		3250			
1·0	+11·000		11 500		2000	$-4a$	
		23 125		5250			
1·5	34·125		16 750		−3250	$6a$	b
		39 875		2000			
2·0	74·000		18 750		3000	$-4a$	$-4b$
		58 625		5000			
2·5	132·625		23 750		−2000	a	$6b$
		82 375		3000			
3·0	215·000		26 750		1000		$-4b$
		109 125		4000			
3·5	324·125		30 750		−250		b
		139 875		3750			
4·0	464·000		34 500		0		
		174 375		3750			
4·5	638·375		38 250				
		212 625					
5·0	851·000						

The fanning out of the incorrect differences indicates that $f(1{\cdot}5)$ and $f(2{\cdot}5)$ contain mistakes. It would also be possible for there to be a mistake in $f(2{\cdot}0)$ if it were not known that there are only two in the table. Assuming that the mistake in $f(1{\cdot}5)$ is a and in $f(2{\cdot}5)$ is b the errors in fourth differences (since fourth differences are zero for an exact cubic) due to these are written down alongisde the fourth differences. The errors due to a are written in a column beginning with the first incorrect fourth difference, those due to b beginning with the last incorrect fourth difference but working backwards as in worked example

5. This gives the incorrect fourth differences as a, $-4a$, $6a+b$, $-4a-4b$, $a+6b$, $-4b$ and b. Comparison with the actual values shows that $a=-500$ and $b=-250$. That is, since the table is given to three decimal places the mistake in $f(1{\cdot}5)$ is $-0{\cdot}5$ and in $f(2{\cdot}5)$ is $-0{\cdot}25$. Thus the true values are $f(1{\cdot}5)=34{\cdot}125-(-0{\cdot}5)=34{\cdot}625$ and $f(2{\cdot}5)=132{\cdot}625-(-0{\cdot}25)=132{\cdot}875$.

5.3.6 Examples

Locate and correct the two mistakes in each of the following tables.

1.

x	0	0·1	0·2	0·3	0·4	0·5	0·6	0·7	0·8	0·9	1·0
$f(x)$	7·000	7·316	7·688	8·152	8·774	9·050	10·456	11·648	13·112	14·884	17·000

cubic

2.

x	0	0·2	0·4	0·6	0·8	1·0	1·2	1·4	1·6	1·8	2·0	2·2	2·4	2·6
$f(x)$	− 5·0000	− 3·9328	− 2.1968	+ 0·4672	4·3952	10·000	17·7112	28·2752	42·1522	60·1312	83·0000	111·6352	146·9872	190·0832

quartic

3.

x	0	0·01	0·02	0·03	0·04	0·05	0·06	0·07	0·08	0·09	0·10
$f(x)$	− 4·0000	− 3·9391	− 3·8764	− 3·8199	− 3·7456	− 3·6775	− 3·6036	− 3·5359	− 3·4624	− 3·3871	− 3·3100

quadratic

5.4 NOTATION FOR DIFFERENCES

It is necessary to be able to refer to a particular difference in a difference table. So that this can be done without having to refer to "the third number in the column of first differences" or by using similar wording a few standard abbreviations are used, some of which follow, others being introduced when necessary later in this book.

The *tabular interval*, i.e. the difference between the successive values of the variable at which the function is tabulated, is denoted by h, provided that the interval is constant.

Different values of the variable are distinguished by labelling one x_0, the values following by $x_1=x_0+h$, $x_2=x_0+2h$, etc. and the values preceding by $x_{-1}=x_0-h$, $x_{-2}=x_0-2h$ etc. so that in general $x_r=x_0+rh$ for $r=0, \pm 1, \pm 2$ etc. The corresponding values of the function are denoted by f_0, f_1, f_2, etc. and f_{-1}, f_{-2}, etc. i.e. $f_0=f(x_0)$, $f_2=f(x_0+2h)$. Thus in general $f_r=f(x_r)=f(x_0+rh)$. Note that the suffixes refer only to relative position of the function values. Usually $f_0 \neq f(0)$, for $f(0)$ means the value of the function at $x=0$, while f_0 means the value at $x=x_0$.

e.g. If in the table

x	0	0·1	0·2	0·3	0·4	0·5	0·6	0·7
$f(x)$	0	0·012	0·072	0·252	0·672	1·500	2·952	5·922

we take 0·3 as x_0 then $x_1 = 0{\cdot}4$, $x_2 = 0{\cdot}5$, $x_3 = 0{\cdot}6$, $x_4 = 0{\cdot}7$ and $x_{-1} = 0{\cdot}2$, $x_{-2} = 0{\cdot}1$, $x_{-3} = 0$, $f_0 = 0{\cdot}252$, $f_1 = 0{\cdot}672$, $f_{-1} = 0{\cdot}072$. Here $h = 0{\cdot}1$.

Differences are denoted by three different symbols, each being used with different formulae and in different circumstances. In §5.5 *forward differences* are used.

The difference between f_1 and f_0, given by $f_1 - f_0$, is denoted by Δf_0 and the difference between f_3 and f_2, given by $f_3 - f_2$, is denoted by Δf_2 i.e. in general:

$$\Delta f_r = f_{r+1} - f_r$$

$$\text{or } \Delta f(x_r) = f(x_r + h) - f(x_r)$$

Δf_0 can be thought of as 'the change in f_0 which gives f_1' since $f_1 = f_0 + \Delta f_0$ by rearranging the definition.

Since second differences are the changes in the first differences it follows that the change in Δf_0 which gives Δf_1 can be represented by $\Delta\Delta f_0$ which is usually written as $\Delta^2 f_0$. Similarly third, fourth and higher differences are represented by $\Delta^3 f$, $\Delta^4 f$, etc.

The relative positions of the forward differences are shown in the following table. Note how the differences with the same suffix slope downwards from the initial function value with that suffix in the direction of increasing values of the variable, hence the term "forward" differences.

⋮	⋮	⋮	⋮	⋮	⋮	⋮
x_{-2}	f_{-2}					⋮
		Δf_{-2}				⋮
x_{-1}	f_{-1}		$\Delta^2 f_{-2}$			⋮
		Δf_{-1}		$\Delta^3 f_{-2}$		⋮
x_0	f_0		$\Delta^2 f_{-1}$		$\Delta^4 f_{-2}$	⋮
		Δf_0		$\Delta^3 f_{-1}$		⋮
x_1	f_1		$\Delta^2 f_0$		$\Delta^4 f_{-1}$	⋮
		Δf_1		$\Delta^3 f_0$		⋮
x_2	f_2		$\Delta^2 f_1$		$\Delta^4 f_0$	⋮
		Δf_2		$\Delta^3 f_1$	⋮	⋮
x_3	f_3		$\Delta^2 f_2$	⋮	⋮	⋮
		Δf_3	⋮	⋮	⋮	⋮
x_4	f_4	⋮	⋮	⋮	⋮	⋮
⋮	⋮	⋮	⋮	⋮	⋮	⋮

5.5 TO DETERMINE THE EQUATION OF A POLYNOMIAL FROM ITS DIFFERENCE TABLE

5.5.1 Connection between differences and degree of polynomial, for unit tabular interval

The reader may have noticed in earlier examples that the columns of constant differences for a quadratic are always multiples of 2(or 2!), for a cubic multiples of 6 (or 3!) and for a quartic multiples of 24 (or 4!) etc. When the tabular interval is unity, the constant differences for $4x^2$ are 8, for $2x^3$ are 12 and for $3x - 5x^2$ are -10. There is a connection between the differences and the degree of the polynomial and also a connection between the differences and the coefficients of the polynomial. In fact the connection is more direct with the coefficients of *factorial polynomials* than with the coefficients of powers of x.

The factorial polynomial of degree n, *with unit interval*, denoted by $[x]^n$ is defined by

$$[x]^n = x(x-1)(x-2)\ldots(x-(n-1))$$

$$\text{e.g. } [x]^3 = x(x-1)(x-2)$$

By definition $[x]^0 = 1$.

For $h = 1$, from the definition of forward differences:

$$\Delta f(x) = f(x+1) - f(x)$$

Hence for $f(x) = [x]^n$ we have

$$\begin{aligned}
\Delta[x]^n &= [x+1]^n - [x]^n \\
&= (x+1)[x]^{n-1} - [x]^{n-1}(x-n+1) \\
&= [x]^{n-1}(x+1-x+n-1)
\end{aligned}$$

$$\text{i.e. } \Delta[x]^n = n[x]^{n-1}$$

$$\begin{aligned}
\text{Also } \Delta^2[x]^n &= \Delta\{\Delta[x]^n\} \\
&= \Delta\{n[x]^{n-1}\} \\
&= n\Delta[x]^{n-1}
\end{aligned}$$

$$\text{i.e. } \Delta^2[x]^n = n(n-1)[x]^{n-2}$$

and similarly for higher differences, giving in general:

$$\Delta^r[x]^n = n(n-1)\ldots(n-r+1)[x]^{n-r}$$

Compare this result with the differentiation of x^n.

5.5.2 Expression of any polynomial in terms of factorial polynomials

We can assume for any polynomial of degree n the form $f(x) = a_0 + a_1[x] + a_2[x]^2 + a_3[x]^3 + \ldots + a_n[x]^n$ in terms of factorial polynomials, since $[x]^n$ is a polynomial of degree n. (Note that all terms except the last one are of a lower degree than n and that the last one is only of degree n).

e.g. The cubic $6x^3 - 12x^2 + 8x + 1$ can be rearranged as

$1 + 2x + 6x(x-1) + 6x(x-1)(x-2)$ which is $1 + 2[x] + 6[x]^2 + 6[x]^3$.

We now wish to obtain expressions for the coefficients $a_0, a_1, \ldots a_n$.

Putting $x = 0$ it follows that $a_0 = f(0)$.

Forming the first difference:

$$\Delta f(x) = \Delta a_0 + \Delta a_1[x] + \Delta a_2[x]^2 + \ldots + \Delta a_n[x]^n$$

hence $\Delta f(x) = a_1 + 2a_2[x] + 3a_3[x]^2 + \ldots + n\,a_n[x]^{n-1}$ and putting $x = 0$ it follows that $a_1 = \Delta f(0)$.

Forming the second difference:

$$\Delta^2 f(x) = 2a_2 + 3.2\, a_3[x] \ldots n(n-1)\, a_n[x]^{n-2}$$

and putting $x = 0$ it follows that $a_2 = [\Delta^2 f(0)]/2$. Similarly expressions for a_3, a_4 etc. can be obtained which are all of the form $a_r = [\Delta^r f(0)]/r!$.

A polynomial can thus be expressed in terms of its differences and factorial polynomials as:

$$f(x) = f(0) + \Delta f(0)\,[x] + \frac{\Delta^2 f(0)}{2!}[x]^2 + \frac{\Delta^3 f(0)}{3!}[x]^3 + \ldots + \frac{\Delta^n f(0)}{n!}[x]^n$$

or on expanding the factorial polynomials:

$$f(x) = f(0) + \Delta f(0)x + \frac{\Delta^2 f(0)}{2!}x(x-1) + \frac{\Delta^3 f(0)}{3!}x(x-1)(x-2) + \ldots$$
$$\ldots + \frac{\Delta^n f(0)}{n!}x(x-1)(x-2)\ldots(x-n+1)$$

5.5.3 Connection between differences and degree of polynomial for non-unit tabular interval

If the polynomial is not tabulated at unit interval but at interval h put

$$z = \frac{x}{h}$$

then $x = hz$

and $f(x) = f(hz)$

which gives $f(x) = F(z)$ say.

Thus $F(0) = f(0)$

and $F(1) = f(h)$

hence $\Delta F(0) = F(1) - F(0)$

$= f(h) - f(0)$

i.e. $\Delta F(0) = \Delta f(0)$

Similarly $\Delta^2 F(0) = \Delta^2 f(0)$ etc.

But $F(z) = F(0) + \Delta F(0)\,[z] + \{\Delta^2 F(0)/2!\}\,[z]^2 + \ldots + \{\Delta^n F(0)/n!\}\,[z]^n$ hence since $F(z) = f(x)$ and $z = x/h$ it follows that

$$f(x) = f(0) + \Delta f(0)\left[\frac{x}{h}\right] + \frac{\Delta^2 f(0)}{2!}\left[\frac{x}{h}\right]^2 + \ldots + \frac{\Delta^n f(0)}{n!}\left[\frac{x}{h}\right]^n$$

or on expanding the factorial polynomials:

$$f(x) = f(0) + \frac{\Delta f(0)}{h}x + \frac{\Delta^2 f(0)}{h^2 2!}x(x-h) + \ldots$$
$$\ldots + \frac{\Delta^n f(0)}{h^n n!}x(x-h)\ldots(x-(n-1)h)$$

5.5.4 Transformation method when tabulation point for x = 0 is missing

If $h = 1$ and if $x = 0$ is not a tabulated point, but the first tabulated point is at $x = x_0$ put $z = x - x_0$ then $x = z + x_0$ and $f(x) = f(z + x_0)$ which gives $f(x) = F(z)$ say.

Thus $F(0) = f(x_0)$ and $F(1) = f(x_1)$

$$\text{hence } \Delta F(0) = F(1) - F(0)$$
$$= f(x_1) - f(x_0)$$
$$\text{i.e. } \Delta F(0) = \Delta f(x_0)$$

Similarly $\Delta^2 F(0) = \Delta^2 f(x_0)$ etc.

$$\text{But } F(z) = F(0) + \Delta F(0)\,[z] + \frac{\Delta^2 F(0)}{2!}\,[z]^2 + \ldots + \frac{\Delta^n F(0)}{n!}\,[z]^n$$

$\therefore$ Since $F(z) = f(x)$ and $z = x - x_0$

$$f(x) = f(x_0) + \Delta f(x_0)\,[x - x_0] + \frac{\Delta^2 f(x_0)}{2!}\,[x - x_0]^2 + \ldots + \frac{\Delta^n f(x_0)}{n!}\,[x - x_0]^n$$

or on expanding the factorial polynomials

$$f(x) = f(x_0) + \Delta f(x_0)\,(x - x_0) + \frac{\Delta^2 f(x_0)}{2!}\,(x - x_0)(x - x_0 - 1) + \ldots$$
$$\ldots + \frac{\Delta^n f(x_0)}{n!}\,(x - x_0) + \ldots + (x - x_0 - (n - 1)).$$

In practice when a polynomial is not tabulated at $x = 0$ or at a unit interval it is usually easier to make a transformation from x into a suitable z and then substitute in the formula of §5.5.2 rather than substitute directly in one of the formulae obtained above. See worked examples 2 and 3 below.

The expansion obtained in §5.5.2 thus makes it possible to obtain the equation of a polynomial when the necessary differences are available.

To find the equation of a polynomial of degree n in this way differences up to the nth must be known. It is therefore necessary to have at least $n + 1$ points tabulated.

Worked example 1

Find the equation of the cubic passing through the points (0, 1), (1, 3), (2, 7) and (3, 19).

On differencing the following table is obtained:

x	$f(x)$			
0	1			
		2		
1	3		2	
		4		6
2	7		8	
		12		
3	19			

Hence $f(0) = 1$, $\Delta f(0) = 2$, $\Delta^2 f(0) = 2$ and $\Delta^3 f(0) = 6$

(Fourth and higher differences are zero since this is a cubic).

$$\therefore f(x) = 1 + 2x + \frac{2}{2}x(x - 1) + \frac{6}{6}x(x - 1)(x - 2)$$

$$= 1 + 2x + x^2 - x + x^3 - 3x^2 + 2x$$

$$f(x) = x^3 - 2x^2 + 3x + 1$$

Check by substitution.

Worked example 2

Find the equation of the polynomial given by the following table:

x	$f(x)$			
0	8·000			
		− 298		
0·1	7·702		12	
		− 286		12
0·2	7·416		24	
		− 262		12
0·3	7·154		36	
		− 226		12
0·4	6·928		48	
		− 178		
0·5	6·750			

Third differences are constant therefore the polynomial is a cubic and fourth and higher differences are zero.

Since the interval of the table is not unity it is necessary to make a transformation. As $h = 1/10$ we put $z = 10x$ as in §5.5.3. i.e. For $x = 0(0{\cdot}1)0{\cdot}5$, $z = 0(1)5$.

Here $f(0) = 8$, $\Delta f(0) = -0{\cdot}298$, $\Delta^2 f(0) = 0{\cdot}012$ and $\Delta^3 f(0) = 0{\cdot}012$.

$$\text{Hence } f(x) = 8 - 0{\cdot}298 \,.\, 10x + \frac{0{\cdot}012}{2} \,.\, 10x(10x - 1) + \frac{0{\cdot}012}{6} \,.\, 10x(10x - 1)(10x - 2)$$

$$= 8 - 2{\cdot}98x + 0{\cdot}06x(10x - 1) + 0{\cdot}02x(100x^2 - 30x + 2)$$

$$= 8 - 2{\cdot}98x + 0{\cdot}6x^2 - 0{\cdot}06x + 2x^3 - 0{\cdot}6x^2 + 0{\cdot}04x$$

i.e. $f(x) = 2x^3 - 3x + 8$

Check by substitution.

Worked example 3.

Find the equation of the cubic passing through the points $(1, 4)$, $(1\frac{1}{2}, 18\frac{1}{4})$, $(2, 44)$ and $(2\frac{1}{2}, 84\frac{1}{4})$.

On differencing the following table is obtained:

x	$f(x)$			
1·0	4·00			
		1425		
1·5	18·25		1150	
		2575		300
2·0	44·00		1450	
		4025		
2·5	84·25			

Since here $h = \frac{1}{2}$ and $x_0 = 1$ the transformation $z = 2(x - 1)$ is made, i.e. For $x = 1{\cdot}0(0{\cdot}5)2.5$, $z = 0(1)3$, $f(x_0) = 4$, $\Delta f(x_0) = 14{\cdot}25$, $\Delta^2 f(x_0) = 11{\cdot}50$ and $\Delta^3 f(x_0) = 3{\cdot}00$.

$$\therefore f(x) = 4 + 14{\cdot}25.2(x - 1) + \frac{11{\cdot}5}{2} \,.\, 2(x - 1)\,[2(x - 1) - 1] + \frac{3}{6} \,.\, 2(x - 1)\,[2(x - 1) - 1]\,[2(x - 1) - 2]$$

$$= 4 + 28{\cdot}5(x - 1) + 11{\cdot}5(x - 1)\,(2x - 3) + (x - 1)\,(2x - 3)(2x - 4)$$

$$= 4 + 28{\cdot}5x - 28{\cdot}5 + 23x^2 - 57{\cdot}5x + 34{\cdot}5 + (x - 1)\,(4x^2 - 14x + 12)$$

$$= 23x^2 - 29x + 10 + 4x^3 - 18x^2 + 26x - 12$$

$$f(x) = 4x^3 + 5x^2 - 3x - 2$$

Check by substitution.

Alternative method when $f(0)$ is not tabulated.

The table may be extended from the differences by the method of §5·2 until $f(0)$ is reached.

For example, the table of example 3 above extended backwards to $f(0)$ is:

x	$f(x)$			
0	−2·00			
		25		
0·5	−1·75		550	
		575		300
1·0	4·00		850	
		1425		300
1·5	18·25		1150	
		2575		300
2·0	44·00		1450	
		4025		
2·5	84·25			

$f(0)$ is now tabulated so that since $h = \frac{1}{2}$ it is only necessary to make the transformation $z = 2x$.

Here $f(0) = -2$, $\Delta f(0) = 0{\cdot}25$, $\Delta^2 f(0) = 5{\cdot}50$ and $\Delta^3 f(0) = 3{\cdot}00$.

$$\text{Hence } f(x) = -2 + 0{\cdot}25.2x + \frac{5{\cdot}5}{2} . 2x(2x-1) + \frac{3}{6} . 2x(2x-1)(2x-2)$$

$$= -2 + 0{\cdot}5\,x + 5{\cdot}5x(2x-1) + x(4x^2 - 6x + 2)$$

$$= -2 + 0{\cdot}5x + 11x^2 - 5{\cdot}5x + 4x^3 - 6x^2 + 2x$$

i.e. $f(x) = 4x^3 + 5x^2 - 3x - 2$.

This second method is used when values of the polynomial near to $x = 0$ are known and it usually makes the terms of the formula easier to simplify.

The theory used in this section is not only applicable to the derivation of the equation of a polynomial from an exact table, but also provides a means of making a polynomial approximation to other functions. For example if a function is tabulated over a small range of values of the variable, and fourth and higher differences are very small and can be neglected, then it will be possible to find a cubic to fit the function closely in the given range.

5.5.5 Examples

Find the equations of the polynomials passing through the following sets of points:

1. (0, − 5) (1, − 10) (2, − 5) (3, 16) (4, 59)

2.

x	0	0·1	0·2	0·3	0·4	0·5
$f(x)$	7	7·164	7·272	7·348	7·416	7·500

3.

x	4	5	6	7	8
$f(x)$	3·6	9·1	17·6	29·7	46·0

4.

x	0·3	0·4	0·5	0·6	0·7
$f(x)$	7·714	7·888	8·050	8·212	8·386

5.

x	0	0·2	0·4	0·6	0·8	1·0
$f(x)$	1·0000	1·5856	2·0976	2·4976	2·7856	3·0000

6.

x	0	0·1	0·2	0·4	0·5	0·6	0·8
$f(x)$	6·000	6·712	7·456	9·088	10·000	10·992	13·264

6

The Solution of Linear Simultaneous Equations

6.1 CHOOSING A SUITABLE TYPE OF SOLUTION

Sets of linear simultaneous equations arise in several different situations, such as in the method of least squares in curve fitting, in surveying, in linear programming or in the application of Kirchoff's law. Likewise there are several different methods for the solution of these sets of equations, the method used depending on the calculating aids available, the type of equations requiring solution and the accuracy required in the solution.

These methods can be divided into two basic types. Firstly direct methods, where the accuracy of the solution depends on the accuracy with which the complete working has been carried out; and secondly iterative methods where the accuracy of the solution depends on the number of times the process is repeated. Two examples of each of these types follow, namely the method of elimination in §6.1 and the method of triangular decomposition in §6.2 as the two direct methods, the method of relaxation in §6.3 and the Gauss-Seidel method in §6.4 as the two indirect methods.

Note that only sets of up to three equations in three unknowns are discussed here, but that all of the methods can be extended for sets of equations with more unknowns.

6.2 THE METHOD OF ELIMINATION

This is the method usually used for the solution of a pair of simultaneous equations, in two unknowns, programmed for use with a calculating machine and incorporating further checks.

For example, consider the solution of the equations:

$$2x + 3y = 8 \qquad (1)$$

$$5x + 4y = 13 \qquad (2)$$

Firstly x is eliminated to obtain y:

Multiply (1) by 5:	$10x + 15y = 40$
Multiply (2) by 2:	$10x + 8y = 26$
Subtract:	$7y = 14$
Divide by 7:	$y = 2$

Secondly this value of y is substituted in one of the original equations to obtain x:

Substitute in (1): $2x + 6 = 8$

thus $2x = 2$

so that $x = 1$

Thirdly the solution is checked in the other equation:

Check in (2): LHS $= 2 + 6$

$= 8$

RHS $= 8$

In the first stage of the solution each equation is multiplied by a different number in order to make the coefficient of x in each equation equal. The same result can be obtained by multiplying just one equation by a suitable number, e.g. in the above example, equation (2) can be multiplied by $2/5 = 0{\cdot}4$ so that the coefficient of x is 2 as in equation (1). Further, taking the coefficients of x and y, and the constant term in turn, the operation of subtraction can be incorporated into one operation on the calculating machine, i.e. $2 - 0{\cdot}4 \times 5 = 0$, $3 - 0{\cdot}4 \times 4 = 1{\cdot}4$ and $8 - 0{\cdot}4 \times 13 = 2{\cdot}8$ giving the equation in y alone: $1{\cdot}4y = 2{\cdot}8$ and thus $y = 2$. If the equations are written down so that the coefficients corresponding to each unknown are in different columns then it is only necessary to write down the unknowns at the top of each column, at the same time omitting the + and = signs. So the working for the first stage of the solution can be written down as follows:

Operation	x	y	c	Equation
	2	3	8	(1)
	5	4	13	(2)
(1) − 0·4 (2)		1·4	2·8	(3)
y		1	2	

The operation column is very necessary when the coefficients include a few significant figures, and there are more unknowns, both to show the value of the multiplier being used and to serve as a reminder of the operation being carried out.

An advantage of this method of solution is that each line of the working can be checked by incorporating a "check sum" column (usually denoted by Σ). The coefficients and constant term of each equation are added together and the operations carried out on the equations are also carried out on these values. The sum of the coefficients in a new equation should be equal to the corresponding value obtained for the check sum. e.g. In the above example Σ for line 1 is $2 + 3 + 8 = 13$, Σ for line 2 is 22. Carrying out the given operation on these values gives $\Sigma = 13 - 0{\cdot}4 \times 22 = 4{\cdot}2$ which is the sum of the coefficients in line 3. These values should be equal since the check sum is derived from $(2 + 3 + 8) - 0{\cdot}4 \times (5 + 4 + 13)$, and the sum of the coefficients on line 3 is derived from $(2 - 0{\cdot}4 \times 5) + (3 - 0{\cdot}4 \times 4) + (8 - 0{\cdot}4 \times 13)$. This working is incorporated into the layout as follows:

Operation	x	y	c	Σ	Equation
	2	3	8	13	(1)
	5	4	13	22	(2)
(1) − 0·4 (2)		1·4	2·8	4·2	(3)
y		1	2	3	

The last line is obtained by dividing line 3 by the coefficient of y, 1·4. Note that the value in the check sum column is again the sum of the coefficients in that line.

The second stage of the solution, called 'back substitution', can also be set out in this tabular form, though no extra checks can be made until the final stage, when the value of the left hand side of one of the original equations is found. The complete working is:

Operation	x	y	c	Σ	Equation
	2	3	8	13	(1)
	5	4	13	22	(2)
(1) − 0·4 (2)		1·4	2·8	4·2$\surd$	(3)
y		1	2	3$\surd$	
Subst. in (1)	2	6	8		
	2		2		
x	1		1		
Check in (2)	LHS = 8 $\surd$				

The complete solution of a set of linear simultaneous equations in any number of unknowns by this method still falls into these three stages. Stage one: elimination; stage two: back substitution; and stage three: the final check.

6.2.1 Choosing multipliers

In the solution of a set of three equations in three unknowns x, y and z, x is first eliminated to give two equations in y and z. Thus two multipliers are used in obtaining these two equations. In order to keep rounding off errors to a minimum the values chosen for the multipliers should always be less than one. Thus the equations

$$a_1x + a_2y + a_3z = a_4 \quad (1)$$

$$b_1x + b_2y + b_3z = b_4 \quad (2)$$

$$c_1x + c_2y + c_3z = c_4 \quad (3$$

should either by written down so that equation (1) has the *smallest* coefficient of x, and this be used in the *numerators* of the multipliers of equations (2) and (3), namely a_1/b_1 and a_1/c_1, or they should be written down so that equation (1) has the *largest* coefficient of x, and this be used in the *denominators* of the multipliers, namely b_1/a_1 and c_1/a_1, which will now both multiply equation (1). The first alternative is used here. There is a similar possibility in obtaining the multiplier used to eliminate y from the two equations in y and z. It is easier to use the same type of multiplier as in eliminating x even when the two equations are not written down in the desired order. See the example below in whch this occurs.

6.2.2 Accuracy of solution

The accuracy attainable in the values of the unknowns depends both on the accuracy with which the coefficients and constant terms are given, and the accuracy of the working. Rounding off errors in the working will not affect the solution provided sufficient guarding figures are used. Two such guarding figures are usually sufficient with small sets of equations, but with larger sets rounding off errors are more likely to accumulate, and more guarding figures may be necessary. In such cases the values obtained for the unknowns may be confirmed by repeating the solution with a further pair of guarding figures, and comparing the two solutions to the required number of figures, repeating this process again if necessary. Errors in the coefficients, or constant terms, or both, limit the accuracy attainable but usually this is nearly the same as the accuracy with which the coefficients or constant terms are given. Such errors are more fully discussed in section 6.3.6. The attainable accuracy may also be restricted if the set of equations to be solved is ill-conditioned (see §6.2.3) but this is not discussed in detail in this book.

Worked example

Find, correct to 2D, the values of x, y and z which satisfy the following equations:

$$4{\cdot}44x - 9{\cdot}94y + 2{\cdot}41z = 5{\cdot}36$$

$$8{\cdot}24x + 2{\cdot}02y - 2{\cdot}18z = 9{\cdot}34$$

$$0{\cdot}93x + 3{\cdot}86y + 11{\cdot}66z = 2{\cdot}57$$

The equations are rewritten in tabular form with the equation having the smallest coefficient of x first, and the check sums for each equation evaluated as follows on the first three lines.

	Operation	x	y	z	c	Σ	Eqn.
1		0·93	3·86	11·66	2·57	19·02	(1)
2		4·44	− 9·94	2·41	5·36	2·27	(2)
3		8·24	2·02	− 2·18	9·34	17·42	(3)

The first stage of the elimination is to obtain a pair of equations in y and z alone by eliminating x. The first such equation (4) is obtained from equations (1) and (2) by the operation: multiply equation (2) by $0{\cdot}93 \div 4{\cdot}44 = 0{\cdot}2095$, and subtract the result from equation (1). e.g. The coefficient of z in equation (4) is obtained by the following operations on the calculating machine:
S.R. 2D   C.R. 4D   Acc. 6D

	S.R.	C.R.	Acc.
Set 11·66 in S.R.	11·66	0	0
Add into Acc. Clear C.R. and S.R.	0	0	11·660 000
Set 2·41 in S.R.	2·41	0	11·660 000
Negative multiply by 0·2095	2·41	0·2095	11·155 105

This result is rounded off to 4D before writing down in the working, i.e. two more places than the number required in the values of the unknowns. It is preferable to obtain the multiplier by tear-down division (§1.15.1) and to apply the operation to the coefficients of x to show that the coefficient of x in equation (4) is zero. This is the only current check on the value of the multiplier. The check sum column does not check this value so that it is possible to use an incorrect multiplier, the effect of which will not be apparent until the final check.

The value in the check sum column is first calculated using the current operation on the appropriate Σ values, (i.e. apply the operation to each column in turn including the check sum column) and then checked by adding up the coefficients on the same line. The second value may differ from the first by one or two in the last place due to rounding off errors, if so, note the different figure for the second value as shown in the grid below on lines 4 and 5 and use this figure in later working.

The second equation in y and z alone, equation (5), is obtained by multiplying equation (3) by $0{\cdot}93 \div 8{\cdot}24 = 0{\cdot}1129$ and subtracting from equation (1).

All these values obtained for equations (4) and (5) are noted in the working as they are calculated, so that the first five lines are now as follows:

	Operation	x	y	z	c	Σ	Eqn.
1		0·93	3·86	11·66	2·57	19·02	(1)
2		4·44	− 9·94	2·41	5·36	2·27	(2)
3		8·24	2·02	− 2·18	9·34	17·42	(3)
4	(1) − 0·2095(2)	0	5·9424	11·1551	1·4471	18·5444$^{(6)}$	(4)
5	(1) − 0·1129(3)	0	3·6319	11·9061	1·5155	17·0533$^{(5)}$	(5)

The next stage in the elimination is to eliminate y between equations (4) and (5) to obtain equation (6) in z alone, from which the value of z is calculated. The multiplier 3·6319 ÷ 5·9424 = 0·6112 is again chosen so as to be less than one, though here it is the second equation which has the smallest coefficient of y. Equation (6), 5·0881 z = 0·6310, gives z = 0·1240 on line 7 on dividing line 6 by 5·0881.

	Operation	x	y	z	c	Σ	Eqn.
1		0·93	3·86	11·66	2·57	19·02	(1)
2		4·44	− 9·94	2·41	5·36	2·27	(2)
3		8·24	2·02	− 2·18	9·34	17·42	(3)
4	(1) − 0·2095 (2)	0	5·9424	11·1551	1·4471	18·5444$^{(6)}$	(4)
5	(1) − 0·1129 (3)	0	3·6319	11·9061	1·5155	17·0533$^{(5)}$	(5)
6	(5) − 0·6112 (4)		0	5·0881	0·6310	5·7190$^{(1)}$	(6)
7	z			1	0·1240	1·1240	

The first stage in the back substitution is to obtain the value of y by substituting this value of z in either equation (4) or equation (5). Here equation (5) is used, and on line 8 in the z column is entered 11·9061z, i.e. 11·9061 × 0·1240 =1·4764 to 4D. This value is then subtracted from that of the constant term giving 3·6319y = 0·0391 on line 9 and thus y = 0·0108 on line 10.

	Operation	x	y	z	c	Σ	Eqn.
1		0·93	3·86	11·66	2·57	19·02	(1)
2		4·44	− 9·94	2·41	5·36	2·27	(2)
3		8·24	2·02	− 2·18	9·34	17·42	(3)
4	(1) − 0·2095 (2)	0	5·9424	11·1551	1·4471	18·5444$^{(6)}$	(4)
5	(1) − 0·1129 (3)	0	3·6319	11·9061	1·5155	17·0533$^{(5)}$	(5)
6	(5) − 0·6112 (4)		0	5·0881	0·6310	5·7190$^{(1)}$	(6)
7	z			1	0·1240	1·1240	
8	Subst. in (5)		3·6319	1·4764	1·5155		
9			3·6319		0·0391		
10	y		1		0·0108		

These values of y and z are now substituted in one of equations (1), (2) and (3) to obtain the value of x. Here equation (3) is used and the values of $2{\cdot}02y$ and $-2{\cdot}18z$ are written on line 11. These are subtracted from the constant term to give $8{\cdot}24x = 9{\cdot}5885$ on line 12, and thus $x = 1{\cdot}1637$ on line 13. Note that lines 8 and 11 would be omitted from the back substitution as the values in the 'c' column on lines 9 and 12 can be obtained on the calculating machine without this intermediate writing down.

The final check is now made by substituting these values of x, y and z in one of the original equations, but not that used in the final stage of the back substitution. Here in equation (1) the left hand side is 2·5698, which checks to the required 2D.

Operation	x	y	z	c	Σ	Eqn.
1	0·93	3·86	11·66	2·57	19·02	(1)
2	4·44	−9·94	2·41	5·36	2·27	(2)
3	8·24	2·02	−2·18	9·34	17·42	(3)
4 (1) − 0·2095 (2)	0	5·9424	11·1551	1·4471	18·5444(6)	(4)
5 (1) − 0·1129 (3)	0	3·6319	11·9061	1·5155	17·0533(5)	(5)
6 (5) − 0·6112 (4)		0	5·0881	0·6310	5·7190(1)	(6)
7 z			1	0·1240	1·1240	
8 Subst. in (5)		3·6319	1·4764	1·5155		
9		3·6319		0·0391		
10 y		1		0·0108		
11 Subst. in (3)	8·24	0·0218	−0·2703	9·34		
12	8·24			9·5885		
13 x	1			1·1637		
14 Check in (1)	LHS = 2·5698					

Thus correct to 2D, $x = 1{\cdot}16$, $y = 0{\cdot}01$, $z = 0{\cdot}12$.

6.2.3 Ill-conditioned equations

Consider the equations

$$x + 1000y = 1 \qquad (1)$$

$$x + 999y = 2 \qquad (2)$$

which have the solution $x = 1{,}001$, $y = -1$. A change of only 0·1 per cent in the value of the coefficient of y in equation (1) gives the equations

$$x + 999y = 1 \qquad (1)$$

$$x + 999y = 2 \qquad (2)$$

which have no solution.

If the coefficient of y in equation (2) is also changed by 0·1 per cent then the equations are:

$$x + 999y = 1 \quad (1)$$

$$x + 1000y = 2 \quad (2)$$

which have the solution $x = -998$, $y = 1$.

Thus only small changes in the values of the coefficients of these equations cause very large changes in the solution. Equations in which this occurs are called "ill-conditioned equations".

If the coefficients of such equations are exact it may be possible to obtain the solution to any desired accuracy, but if the coefficients have been rounded off then the attainable accuracy will be much less than that with which the coefficients are given. This accuracy cannot be improved unless more significant figures may be obtained for the coefficients.

The fact that a set of equations is ill-conditioned will often be shown in the above method by a loss of significant figures in the coefficients of the equations obtained in the elimination.

6.3 THE METHOD OF TRIANGULAR DECOMPOSITION

A knowledge of the properties of matrices is assumed in this section.

6.3.1 Solution by using a square matrix property

This method depends on the fact that a square matrix can be expressed as the product of a lower triangular matrix and an upper triangular matrix.

Let the equations to be solved be:

$$a_{11}x_1 + a_{12}x_2 + a_{13}x_3 = c_1$$

$$a_{21}x_1 + a_{22}x_2 + a_{23}x_3 = c_2$$

$$a_{31}x_1 + a_{32}x_2 + a_{33}x_3 = c_3$$

and let $\mathbf{A}$ be the matrix of coefficients, $\mathbf{x}$ the column vector (x_1, x_2, x_3) and $\mathbf{c}$ the column vector (c_1, c_2, c_3). Then the equations can be written in the form:

$$\mathbf{Ax} = \mathbf{c} \quad (1)$$

The square matrix $\mathbf{A}$ can be decomposed into two triangular matrices: $\mathbf{L}$ a lower triangular matrix, (i.e. a matrix in which all the elements above the diagonal are zero) and $\mathbf{U}$ an upper triangular matrix (i.e. a matrix in which all the elements below the diagonal are zero).

$$\text{i.e. } \mathbf{A} = \mathbf{L}\,\mathbf{U} \quad (2)$$

Thus $$\mathbf{L}\,\mathbf{U}\,\mathbf{x} = \mathbf{c}$$

and putting $$\mathbf{U}\,\mathbf{x} = \mathbf{y} \quad (3)$$

gives $$\mathbf{L}\,\mathbf{y} = \mathbf{c} \quad (4)$$

As will be seen below **y** can be obtained from equations (4) by successive substitutions, and is then substituted in equations (3) to give **x**, again by successive substitutions. These successive substitutions are possible as **L** and **U** are triangular matrices.

6.3.2 Stages of a solution

In practice the method falls into three main stages, whatever the number of equations. Stage one: the decomposition of matrix **A** into matrices **L** and **U**. Stage two: the determination of **y**. Stage three: the determination of **x**.

Stage one: To obtain **L** and **U**

The diagonal elements of either **L** or **U** can be chosen at random, the values of the other elements then being determined from these. It is usually easiest to take the diagonal elements of **L** as unity.

Thus $\mathbf{L} = \begin{pmatrix} 1 & 0 & 0 \\ l_{21} & 1 & 0 \\ l_{31} & l_{32} & 1 \end{pmatrix}$ and $\mathbf{U} = \begin{pmatrix} u_{11} & u_{12} & u_{13} \\ 0 & u_{22} & u_{23} \\ 0 & 0 & u_{33} \end{pmatrix}$

and we need $\mathbf{L}\,\mathbf{U} = \mathbf{A}$, that is:

$$\begin{pmatrix} 1 & 0 & 0 \\ l_{21} & 1 & 0 \\ l_{31} & l_{32} & 1 \end{pmatrix} \begin{pmatrix} u_{11} & u_{12} & u_{13} \\ 0 & u_{22} & u_{23} \\ 0 & 0 & u_{33} \end{pmatrix} = \begin{pmatrix} a_{11} & a_{12} & a_{13} \\ a_{21} & a_{22} & a_{23} \\ a_{31} & a_{32} & a_{33} \end{pmatrix}$$

thus giving nine equations which are solved in the following order for the nine unknowns in **L** and **U**, proceeding from left to right along each row of **A** in turn. $r_1(\mathbf{L})c_1(\mathbf{U}) = a_{11}$*, $r_1(\mathbf{L})c_2(\mathbf{U}) = a_{12}$ and $r_1(\mathbf{L})c_3(\mathbf{U}) = a_{13}$ give the equations $u_{11} = a_{11}$, $u_{12} = a_{12}$, $u_{13} = a_{13}$ respectively, which determine the values of the elements of the first row of **U**. $r_2(\mathbf{L})c_1(\mathbf{U}) = a_{21}$, i.e. $l_{21}\ u_{11} = a_{21}$ gives the value of l_{21}. $r_2(\mathbf{L})c_2(\mathbf{U}) = a_{22}$ and $r_2(\mathbf{L})c_3(\mathbf{U}) = a_{23}$ give the equations $l_{21}\ u_{12} + u_{22} = a_{22}$ and $l_{21}\ u_{13} + u_{23} = a_{23}$ respectively, which determine u_{22} and u_{23}.

$r_3(\mathbf{L})c_1(\mathbf{U}) = a_{31}$, i.e. $l_{31}\,u_{11} = a_{31}$ gives the value of l_{31}, and $r_3(\mathbf{L})c_2(\mathbf{U}) = a_{32}$, i.e. $l_{31}\,u_{12} + l_{32}\,u_{22} = a_{32}$ gives the value of l_{32}.

Finally, $r_3(\mathbf{L})c_3(\mathbf{U}) = a_{33}$, i.e. $l_{31}\ u_{13} + l_{32}\ u_{23} + u_{33} = a_{33}$ gives the value of u_{33}.

Note how the rows of **U** and **L** are obtained alternately and that division is only needed to find the three unknown elements of **L**.

Stage 2: To obtain **y**

Writing equations (4), $\mathbf{L}\,\mathbf{y} = \mathbf{c}$, in full we have:

$$\begin{pmatrix} 1 & 0 & 0 \\ l_{21} & 1 & 0 \\ l_{31} & l_{32} & 1 \end{pmatrix} \begin{pmatrix} y_1 \\ y_2 \\ y_3 \end{pmatrix} = \begin{pmatrix} c_1 \\ c_2 \\ c_3 \end{pmatrix}$$

* r_i (**L**)c_j (**U**) denotes the product of the ith row of **L** and the jth column of **U**.

giving the three equations:

$$\begin{aligned} y_1 &= c_1 \\ l_{21}\, y_1 + y_2 &= c_2 \\ l_{31}\, y_1 + l_{32}\, y_2 + y_3 &= c_3 \end{aligned}$$

from which y_1, y_2 and y_3 may be calculated by successive substitutions.

Stage 3: To obtain **x**

Writing equations (3), $\mathbf{U}\,\mathbf{x} = \mathbf{y}$, in full we have:

$$\begin{pmatrix} u_{11} & u_{12} & u_{13} \\ 0 & u_{22} & u_{23} \\ 0 & 0 & u_{33} \end{pmatrix} \begin{pmatrix} x_1 \\ x_2 \\ x_3 \end{pmatrix} = \begin{pmatrix} y_1 \\ y_2 \\ y_3 \end{pmatrix}$$

giving the three equations:

$$\begin{aligned} u_{11}\, x_1 + u_{12}\, x_2 + u_{13}\, x_3 &= y_1 \\ u_{22}\, x_2 + u_{23}\, x_3 &= y_2 \\ u_{33}\, x_3 &= y_3 \end{aligned}$$

from which x_3, x_2 and x_1 may be calculated by successive substitutions.

6.3.3 Layout and checks

One method of writing down the necessary working, incorporating checks, is as follows: (Other methods differing slightly from this one are also used*).

$$\begin{array}{llllll} & \mathbf{A} & & & \mathbf{c} & \boldsymbol{\Sigma} \\ a_{11} & a_{12} & a_{13} & & c_1 & \Sigma_1 \\ a_{21} & a_{22} & a_{23} & & c_2 & \Sigma_2 \\ a_{31} & a_{32} & a_{33} & & c_3 & \Sigma_3 \\ & \mathbf{L} & & & & \\ 1 & 0 & 0 & & & \\ l_{21} & 0 & 0 & & & \\ l_{31} & l_{32} & 1 & & & \\ & \mathbf{U} & & & \mathbf{y} & \mathbf{S} \\ u_{11} & u_{12} & u_{13} & & y_1 & S_1 \\ 0 & u_{22} & u_{23} & & y_2 & S_2 \\ 0 & 0 & u_{33} & & y_3 & S_3 \\ \mathbf{x} & & x_3 & x_2 & & x_1 \end{array}$$

Σ_1, Σ_2 and Σ_3 are the sums of the coefficients and constant term of each of the original equations, i.e. the sums of the elements of **A** and of **c** on the same line in the scheme above. Similarly S_1, S_2 and S_3 are the sums of the elements of **U** and **y** on the corresponding lines.

* e.g. See Hartree pp. 180–184 for a method which involves more writing down.

If for the matrices **A**, **B**, **C**, **D**, **E**, **F**, and **G**, **A B** = **C**, **A D** = **E** and **F** and **G** are the column vectors formed by adding together the elements in the corresponding rows of **B**, **D** and **C**, **E** respectively, then it can be shown on expansion and multiplication that **A F** = **G**. Thus here since **L U** = **A** and **L y** = **c**, then **L S** = **Σ** i.e.

$$S_1 = \Sigma_1$$

$$l_{21} S_1 + S_2 = \Sigma_2$$

and
$$l_{31} S_1 + l_{32} S_2 + S_3 = \Sigma_3$$

which provides a check on the values obtained in **L**, **U** and **y**.

Worked example

Find, correct to 2D, the values of x_1, x_2 and x_3 which satisfy the following equations:

$$4{\cdot}44x_1 - 9{\cdot}94x_2 + 2{\cdot}41x_3 = 5{\cdot}36$$

$$8{\cdot}24x_1 + 2{\cdot}02x_2 - 2{\cdot}18x_3 = 9{\cdot}34$$

$$0{\cdot}93x_1 + 3{\cdot}86x_2 + 11{\cdot}66x_3 = 2{\cdot}57$$

The equation with the largest coefficient of x_1 is written down first so that the values of l_{21} and l_{31} are less than one, thus keeping rounding off errors in u_{22} and u_{23} to a minimum. It is desirable that all the elements of **L** should be less than one, just as all the multipliers are chosen in the method of elimination, but this is often difficult to achieve without adding complications to the method, especially with only a hand calculating machine available.

The working is carried out as follows to 4D, since it is desirable to carry two guarding figures, as for the method of elimination, As described above **L** and **U** are found together, then **y**, and finally **x**.

	A		**c**	**Σ**
8·24	2·02	−2·18	9·34	17·42
4·44	− 9·94	2·41	5·36	2·27
0·93	3·86	11·66	2·57	19·02

	L	
1	0	0
0·5388	1	0
0·1129	− 0·3293	1

	U		**y**	**S**
8·24	2·02	− 2·18	9·34	17·42
0	− 11·0284	3·5846	0·3276	− 7·1162
0	0	13·0865	1·6234	14·7099

$x_1 = 0{\cdot}1241$ $\quad x_2 = 0{\cdot}0106$ $\quad x_1 = 1{\cdot}1637$

Hence, correct to 2D, $x_1 = 1{\cdot}16$, $x_2 = 0{\cdot}01$, $x_3 = 0{\cdot}12$.

In the method given here no check on the working is made until after **L**, **U** and **y** have been calculated completely, as the working is easier to follow in this order. If an earlier check is desired then y_2 and S_2 may be obtained as soon as the first two rows of **L** and **U** are known. As in the checks in the method of elimination the pairs of values obtained for S_2 and S_3 may not be exactly equal but may differ by one or two in the last figure.

A final check should be made by substituting the values obtained for x_1, x_2 and x_3 in the original equations.

6.3.4 Symmetric matrices

The matrix of coefficients of the equations obtained in the least squares method of curve fitting is symmetrical about the leading diagonal. There are other situations in which such symmetric matrices arise naturally and in such cases, instead of calculating **L** with the elements of its leading diagonal all unity, it may be better to use the following method. A symmetric square matrix can be expressed as the product of a lower triangular matrix and an upper triangular matrix, each of which is the transpose of the other, so that here **A** can be expressed as $\mathbf{L}\,\mathbf{L}'$ which gives a set of six equations, from which the elements of **L** may be found. The elements of the leading diagonal of **L** are now the square roots of numbers, which may cause difficulty if the elements of **A** are such that the square roots of negative numbers have to be found. This method will not be considered further here.

6.3.5 Comparison with the method of elimination

The method of elimination is easier to follow and is probably the better method to use when sets of linear simultaneous equations are only solved occasionally. Also in the method of elimination checking takes place at an earlier stage of the calculation. On the other hand those who frequently solve sets of equations by hand methods often have a personal preference for this type of method. It also forms the basis of many computer methods, which are extensively used. Further the working in the method of elimination is not used in calculating the attainable accuracy, while the working for the method of triangular decomposition can be used for this as is explained below.

6.3.6 Attainable accuracy

Let the errors in **A** be $\delta\mathbf{A}$ (i.e. any coefficient a_{ij} is in error by δa_{ij}) and in **c** be $\delta\mathbf{c}$, then the corresponding errors in **x**, $\delta\mathbf{x}$ are given as follows in the various possible cases.

(i) **A** exact but **c** not.

Then the exact set of equations, of which $\mathbf{A}\,\mathbf{x} = \mathbf{c}$ is an approximation, is $\mathbf{A}(\mathbf{x} + \delta\mathbf{x}) = \mathbf{c} + \delta\mathbf{c}$

thus $$\mathbf{A}\,\delta\mathbf{x} = \delta\mathbf{c}$$

and hence $$|\,\delta\mathbf{x}\,| \leqslant |\,\mathbf{A}^{-1}\,|\;|\,\delta\mathbf{c}\,|^*$$

* In this section $|\,\mathbf{A}\,|$ does not mean the determinant of **A** but denotes the matrix whose elements are the moduli of the corresponding elements of **A**.

(ii) **c** exact but **A** not.

Then the exact set of equations is:

$$(\mathbf{A} + \delta\mathbf{A})(\mathbf{x} + \delta\mathbf{x}) = \mathbf{c}$$

thus $$\mathbf{A}\,\delta\mathbf{x} + \delta\mathbf{A}\,\mathbf{x} = 0 \text{ neglecting } \delta\,\mathbf{A}\,\delta\mathbf{x}$$

and hence $$|\,\delta\mathbf{x}\,| \leqslant |\,\mathbf{A}^{-1}\,|\,|\,\delta\mathbf{A}\,|\,|\,\mathbf{x}\,|$$

(iii) Neither **A** nor **c** is exact.

From (i) and (ii)

$$|\,\delta\mathbf{x}\,| \leqslant |\,\mathbf{A}^{-1}\,|\,|\,\delta\mathbf{c}\,| + |\,\mathbf{A}^{-1}\,|\,|\,\delta\mathbf{A}\,|\,|\,\mathbf{x}\,|$$

The main difficulty in calculating the errors in **x** is the fact that $\mathbf{A}^{-1}$ must be known in each of the above cases. The calculation of $\mathbf{A}^{-1}$ involves as least as much computation as the solution of the equations, but it can be found fairly quickly when the method of triangular decomposition is used, since $\mathbf{A}^{-1} = \mathbf{U}^{-1}\mathbf{L}^{-1}$, and the inverse of a triangular matrix is much easier to calculate than the inverse of a square matrix. (This method is the basis of a method of computing the inverse of a square matrix.)

Here $$\mathbf{L}^{-1} = \begin{pmatrix} 1 & 0 & 0 \\ -l_{21} & 1 & 0 \\ l_{21}l_{32} - l_{31} & -l_{32} & 1 \end{pmatrix}$$

and $$\mathbf{U}^{-1} = \begin{pmatrix} \dfrac{u_{22}u_{33}}{d} & \dfrac{-u_{12}u_{33}}{d} & \dfrac{u_{12}u_{23} - u_{22}u_{13}}{d} \\ 0 & \dfrac{u_{11}u_{13}}{d} & -\dfrac{u_{11}u_{23}}{d} \\ 0 & 0 & \dfrac{u_{11}u_{12}}{d} \end{pmatrix}$$

where $d = u_{11}\,u_{22}\,u_{33}$.

Usually the elements of $\delta\mathbf{A}$ or $\delta\mathbf{c}$, or both, will not be known exactly but will have some maximum limit from rounding off errors, which thus gives maximum values for the errors in the values of the unknowns. E.g. If in the previous worked example, § 6.3.3 the values of the coefficients and constant terms are rounded off to 2D then $|\,\delta a_{ij}\,| \leqslant 0{\cdot}005$ and $|\,\delta \mathbf{c}_i\,| \leqslant 0{\cdot}005$.

In this example $$\mathbf{L}^{-1} = \begin{pmatrix} 1 & 0 & 0 \\ -0{\cdot}5388 & 1 & 0 \\ -0{\cdot}2903 & 0{\cdot}3293 & 1 \end{pmatrix}$$

and $$\mathbf{U}^{-1} = \begin{pmatrix} 0{\cdot}1214 & 0{\cdot}0222 & -0{\cdot}0141 \\ 0 & -0{\cdot}0907 & 0{\cdot}0248 \\ 0 & 0 & 0{\cdot}0764 \end{pmatrix}$$

hence $$\mathbf{A}^{-1} = \begin{pmatrix} 0{\cdot}1135 & 0{\cdot}0176 & -0{\cdot}0141 \\ 0{\cdot}0417 & -0{\cdot}0825 & 0{\cdot}0248 \\ -0{\cdot}0222 & 0{\cdot}0252 & 0{\cdot}0764 \end{pmatrix}$$

also $|\delta\mathbf{A}| \leqslant \begin{pmatrix} 0{\cdot}005 & 0{\cdot}005 & 0{\cdot}005 \\ 0{\cdot}005 & 0{\cdot}005 & 0{\cdot}005 \\ 0{\cdot}005 & 0{\cdot}005 & 0{\cdot}005 \end{pmatrix}$ and $|\delta\mathbf{c}| \leqslant \begin{pmatrix} 0{\cdot}005 \\ 0{\cdot}005 \\ 0{\cdot}005 \end{pmatrix}$

thus $|\mathbf{A}^{-1}|\,|\delta\mathbf{c}| \leqslant \begin{pmatrix} 0{\cdot}0007 \\ 0{\cdot}0007 \\ 0{\cdot}0006 \end{pmatrix}$ and $|\mathbf{A}^{-1}|\,|\delta\mathbf{A}|\,|\mathbf{x}| \leqslant \begin{pmatrix} 0{\cdot}0009 \\ 0{\cdot}0010 \\ 0{\cdot}0008 \end{pmatrix}$

so that $|\delta\mathbf{x}| \leqslant |\mathbf{A}^{-1}|\,|\delta\mathbf{c}| + |\mathbf{A}^{-1}|\,|\delta\mathbf{A}|\,|\mathbf{x}|$

$$= \begin{pmatrix} 0{\cdot}0016 \\ 0{\cdot}0017 \\ 0{\cdot}0014 \end{pmatrix}$$

Hence the values of the unknowns may be obtained correctly rounded off to 2D or given to 3D with an error of up to two in the last place.

6.4 THE METHOD OF RELAXATION

6.4.1 Operations and residuals showing how to reduce residuals to the required accuracy:

Worked example 1

Consider the set of equations:

$$3x + y = 5$$
$$x + 5y = 7$$

which have the solution $x = 1\frac{2}{7}$, $y = 1\frac{1}{7}$.

If x', y' are approximate values of x and y then $R_1 = 3x' + y' - 5$ and $R_2 = x' + 5y' - 7$ are called the *residuals*. If x' and y' are the exact values of x and y then R_1 and R_2 will both be zero. Usually however we require a solution to a limited accuracy, in which case we seek values of x' and y' so that the residuals will be negligible within this limit.

Values of x' and y' are found by first choosing approximations and calculating the residuals for these. The magnitudes of the residuals are then reduced by suitable changes in these initial values of x' and y'. In order to make such changes it is necessary to know the effect on the residuals of any change in the values of x' and y'. This in turn depends on the coefficients of the equations. For example, a change of 1 in x causes a change of 3 in R_1 and 1 in R_2. A change of -2 in y causes a change of -2 in R_1 and -10 in R_2. For ease of reference during the working it is convenient to summarise the effects on the residuals of unit changes in the unknowns in a table, usually referred to as the *operations table*, each change in value of an unknown being described as an *operation*. In this example the operations table is as follows:

Operation No.	δx	δy	δR_1	δR_2
(i)	1		3	1
(ii)		1	1	5

Integral change in value of unknowns

If no initial approximations to the values of the unknowns are available then the solution is begun with these all zero. The initial values of the residuals are then − 5 and − 7, and these facts are noted on line 1 of the working below.

	Operation	x	y	R_1	R_2
1		0	0	− 5	− 7
2	+ (ii)		1	− 4	− 2
3	+ (i)	1		− 1	− 1
4		1	1	√	√

Operation (ii) is now used to reduce R_2. This gives the new values of the residuals as shown on line 2, namely $-5+1=-4$ and $-7+5=-2$. The operation used and its effect on y are also noted on line 2 in the appropriate columns. R_1 may now be reduced by a change of 1 in x, using operation (i), which gives $R_1=-4+3=-1$ and $R_2=-2+1=-1$ as shown on line 3. No other *integral* change in the values of the unknowns reduces the residuals further so that we now have the approximations $x=1$, $y=1$. These are noted on line 4 and substituted in the original equations to check the values of the residuals.

Choice of operation order

The above operations were not chosen by trial and error but as R_2 was initially the residual of largest magnitude an operation to reduce this was sought. *Usually the operation which reduces the magnitude of the largest residual is used at any stage of the working.*

Non-integral changes in values of unknowns

(a) Changes in the *first* decimal place of each of the unknowns are now considered. Here $R_1=-1{\cdot}0$ and $R_2=-1{\cdot}0$ are noted on line 5 below and as these are both of the same magnitude either may be reduced. In this case R_2 is eliminated by a change of 0·2 in y, i.e. using operation (ii), which is shown on line 6. R_1 is now greatest in magnitude and is reduced using operation (i) as shown on line 7 and finally R_2 is reduced on line 8. No further changes in the first decimal place reduce the residuals, so that the new approximations to the values of x and y are now noted on line 9 and then substituted in the original equation to check the current values of the residuals.

	Operation	x	y	R_1	R_2
1		0	0	− 5	− 7
2	+ (ii)		1	− 4	− 2
3	+ (i)	1		− 1	− 1
4		1	1	√	√
5				− 1·0	− 1·0
6	+ 0·2 (ii)		0·2	− 0·8	0
7	+ 0·3 (i)	0·3		0·1	0·3
8	− 0·1 (ii)		− 0·1	0	− 0·2
9		1·3	1·1	√	√

It will now be noticed that every number on lines 5 to 8 is in terms of the first decimal place. Because of this it is unnecessary to include the decimal point on these lines, and in fact this also makes the working easier as it is always in terms of integers. The revised working is now set out as follows:

	Operation	x	y	R_1	R_2
1		0	0	− 5	− 7
2	+ (ii)		1	− 4	− 2
3	+ (i)	1		− 1	− 1
4		1	1	√	√
5				− 10	− 10
6	+ 2 (ii)		2	− 8	0
7	+ 3 (i)	3		1	3
8	− (ii)		− 1	0	− 2
9		1·3	1·1	√	√

The decimal place must be shown in the values of the unknowns on the check line 9, and must be borne in mind when the check is made.

(b) Changes in the *second* decimal place of each of the unknowns are now considered. In practice this means multiplying the values of the residuals shown on line 8 by 10, as noted on line 10 below, and reducing these as far as possible, as shown on lines 11 and 12.

(c) The process is repeated on lines 14 to 17 where the working is now in terms of the *third* decimal place.

	Operation	x	y	R_1	R_2
1		0	0	− 5	− 7
2	+ (ii)		1	− 4	− 2
3	+ (i)	1		− 1	− 1
4		1	1	√	√
5				− 10	− 10
6	+ 2 (ii)		2	− 8	− 0
7	+ 3 (i)	3		1	3
8	− (ii)		− 1	0	− 2
9		1·3	1·1	√	√
10				0	− 20
11	+ 4 (ii)		4	4	0
12	− (i)	− 1		1	− 1
13		1·29	1·14	√	√
14				10	− 10
15	+ 2 (ii)		2	12	0
16	− 4 (i)	− 4		0	− 4
17	+ (ii)		1	1	1
18		1·286	1·143	√	√

Accuracy considerations

In iterative methods for the solution of linear simultaneous equations, as in other iterative methods for the solution of equations, the process should be continued until two successive approximations agree to the required number of figures. Thus in the above example, as the last two approximations agree to two decimal places, we have $x = 1{\cdot}29$ and $y = 1{\cdot}14$, correct to 2D.

The attainable accuracy is limited as before by rounding-off errors in the coefficients and constants, and in ill-conditioned sets of equations.

It will be seen that one advantage of this method is that most of the working can be done mentally, only the checking of residuals being carried out on a calculating machine. Also if it is desired to obtain a solution of greater accuracy, then the process is simply continued from the stage reached, and a complete reworking is not necessary as it would be with a direct method.

Operations for optimum change in single residuals

In the above example a unit change in x caused a change of 3 in R_1 and 1 in R_2, i.e. the operation (i) was useful in reducing R_1 without having a large effect on R_2. Similarly operation (ii) was useful in reducing R_2 without a great effect on R_1. This type of operation which causes the greatest change in only one residual is the most useful, and it is desirable to have one such operation for each residual.

Consider the equations:

$$a_{11}x + a_{12}y + a_{13}z = c_1$$
$$a_{21}x + a_{22}y + a_{23}z = c_2$$
$$a_{31}x + a_{32}y + a_{33}z = c_3$$

for which the unit operations table is:

No.	δx	δy	δz	δR_1	δR_2	δR_3
(i)	1			a_{11}	a_{21}	a_{31}
(ii)		1		a_{12}	a_{22}	a_{32}
(iii)			1	a_{13}	a_{23}	a_{33}

Operation (i) will affect R_1 more than R_2 and R_3 if

$$d_1 = \frac{|a_{21}| + |a_{31}|}{|a_{11}|} < 1.$$

Similarly operations (ii) and (iii) will have the greatest effect on R_2 and R_3 if

$$d_2 = \frac{|a_{12}| + |a_{32}|}{|a_{22}|} < 1 \quad \text{and} \quad d_3 = \frac{|a_{13}| + |a_{23}|}{|a_{33}|} < 1$$

respectively. In the above example $d_1 = \frac{1}{3}$ and $d_2 = \frac{1}{5}$ indicate that good convergence was to be expected.

In General all sets of equations which have a *dominant leading diagonal* are most suitable for solution by this method.

Worked example 2.

Solve, correct to 2D, the equations

$$4x - y + z = 3$$
$$2x + 5y - z = 2$$
$$x + 2y + 6z = 5$$

The residuals are given by:

$$R_1 = 4x - y + z - 3$$
$$R_2 = 2x + 5y - z - 2$$
$$R_3 = x + 2y + 6z - 5$$

and the operations table is:

No.	δx	δy	δz	δR_1	δR_2	δR_3	d
(i)	1			4	2	1	$d_1 = \frac{3}{4}$
(ii)		1		-1	5	2	$d_2 = \frac{3}{5}$
(iii)			1	1	-1	6	$d_3 = \frac{1}{3}$

No initial approximations to the values of the unknowns are available so that zero values are used, for which the residuals are $R_1 = -3$, $R_2 = -2$ and $R_3 = -5$ as noted on line 1 below.

Since R_3 is the largest residual, operation (iii), which affects R_3 more than R_1 and R_2, is used to reduce it and gives new values of the residuals as shown on line 2, namely $-3 + 1 = -2$, $-2 - 1 = -3$ and $-5 + 6 = 1$.

Although R_2 is now the largest residual, operation (ii) gives new residuals -3, 2 and 3 and thus does not appreciably reduce all the residuals. However operation (i) does give a slight overall reduction in the magnitude of the residuals, and so is used here on line 3.

No further operation reduces the residuals, so the current approximations to the values of the unknowns are noted on line 4, and substituted in the original equations to check the current values of the residuals.

	Operation	x	y	z	R_1	R_2	R_3
1		0	0	0	-3	-2	-5
2	+(iii)			1	-2	-3	1
3	+(i)	1			2	-1	2
4		1	0	1	$\surd$	$\surd$	$\surd$

The residuals are now multiplied by 10 and written on line 5, so that the working is now in terms of the first decimal place.

The process of reducing the residuals by reducing the largest at any stage is continued on lines 6 to 10 where again no operation has an appreciable affect on their magnitude. The changes in values of the unknowns shown on lines 6 to 10 are in terms of the first decimal place, so that, for example, the resulting change in the value of x of -3 gives the next approximation to x of $1{\cdot}0 - 0{\cdot}3 = 0{\cdot}7$, and similarly for y and z as noted on line 11.

The current values of the residuals are again checked using these new approximations. The true values of the residuals at this stage are 0·1, 0·3 and $-0{\cdot}1$ since the working is in terms of the first decimal place, and these will be the values obtained on substituting $x = 0{\cdot}7$, $y = 0{\cdot}3$ and $z = 0{\cdot}6$ in the original equations.

	Operation	x	y	z	R_1	R_2	R_3
5					20	−10	20
6	−5 × (i)	−5			0	−20	15
7	+4 × (ii)		4		−4	0	23
8	−4 × (iii)			−4	−8	4	−1
9	+2 × (i)	2			0	8	1
10	− (ii)		−1		1	3	−1
11		0·7	0·3	0·6	√	√	√

The process is repeated on lines 12 to 30 as shown below, it being necessary to continue to four decimal places here, as the approximations to x and y do not agree to two decimal places on lines 17 and 24.

Thus, correct to 2D, $x = 0{\cdot}66$, $y = 0{\cdot}26$ and $z = 0{\cdot}64$. Note however that the solution has also been obtained correct to 3D, as the approximations on lines 24 and 30 agree to this number of places.

	Operation	x	y	z	R_1	R_2	R_3
1		0	0	0	−3	−2	−5
2	+(iii)			1	−2	−3	1
3	+(i)	1			2	−1	2
4		1	0	1	√	√	√
5					20	−10	20
6	− 5 × (i)	−5			0	−20	15
7	+ 4 × (ii)		4		−4	0	23
8	− 4 × (iii)			−4	−8	4	−1
9	+ 2 × (i)	2			0	8	1
10	− (ii)		−1		1	3	−1
11		0·7	0·3	0·6	√	√	√
12					10	30	−10
13	− 6 × (ii)		−6		16	0	−22
14	+ 4 × (iii)			4	20	−4	2
15	− 5 × (i)	−5			0	−14	−3
16	+ 3 × (ii)		3		−3	1	3
17		0·65	0·27	0·64	√	√	√

	Operation	x	y	z	R_1	R_2	R_3
18					−30	10	30
19	$-5 \times$ (iii)			−5	−35	15	0
20	$+9 \times$ (i)	9			1	33	9
21	$-6 \times$ (ii)		−6		7	3	−3
22	$-2 \times$ (i)	−2			−1	−1	−5
23	+ (iii)			1	0	−2	1
24		0·657	0·264	0·636	√	√	√
25					0	−20	10
26	$+4 \times$ (ii)		4		−4	0	18
27	$-3 \times$ (iii)			−3	−7	3	0
28	$+2 \times$ (i)	2			1	7	2
29	− (ii)		−1		2	2	0
30		0·6572	0·2643	0·6357	√	√	√

so that correct to 2D, $x = 0{\cdot}66$, $y = 0{\cdot}26$, $z = 0{\cdot}64$.

Examples for exercise are to be found in questions 5, 6 and 7 of §6.6.

Non-integral coefficients

In the previous examples the coefficients were all small integers and the working was easily carried out exactly. If however the coefficients are not small integers then much work is involved if this is attempted. One method which may be used to simplify the working is to round off the coefficients used in the operations table, but to calculate the residuals exactly at each checking stage and to work from these values in subsequent lines. This means that the residuals will only rarely check exactly at the checking stage, and that if the exactly calculated residuals are much larger than the approximiate residuals, it will be necessary to reduce them further before carrying on to the next decimal place. Alternatively initial approximations to the values of the unknowns may be found using the rounded off operations table, which are then used to start a solution with the exact operations table. The first method is usually quicker and thus preferable.

Worked example 3.

Find, correct to 2D, the values of x, y and z which satisfy the equations:

$$8{\cdot}24x + 2{\cdot}02y - 2{\cdot}18z = 9{\cdot}34$$

$$4{\cdot}44x - 9{\cdot}94y + 2{\cdot}41z = 5{\cdot}36$$

$$0{\cdot}93x + 3{\cdot}86y + 11{\cdot}66z = 2{\cdot}57$$

Operations table: (the coefficients are here rounded off to the nearest integer.)

No.	δx	δy	δz	δR_1	δR_2	δR_3	
(i)	1			8	4	1	$d_1 = \frac{5}{8}$
(ii)		1		2	−10	4	$d_2 = \frac{3}{5}$
(iii)			1	−2	2	12	$d_3 = \frac{1}{3}$

The first approximation to each of the unknowns is again taken as zero, but the residuals − 9·34, − 5·36 and − 2·57 are rounded off to the nearest integer before writing on line 1 of the working as follows:

	Operation	x	y	z	R_1	R_2	R_3
1		0	0	0	−9	−5	−3
2	+ (i)	1			−1	−1	−2
3		1	0	0	−1·10	−0·92	−1·64
4					−11	−9	−16
5	+ (iii)			1	−13	−7	−4
6	+ 2 × (i)	2			3	1	−2
7		1·2	0·0	0·1	3·30	2·09	−2·88
8					33	21	−29
9	− 4 × (i)	−4			1	5	−33
10	+ 2 × (iii)			2	−3	9	−9
11	+ (ii)		1		−1	−1	−5
12		1·16	0·01	0·12	−2·30	−1·98	−5·34
13					−23	−20	−53
14	+ 4 × (iii)			4	−31	−12	−5
15	+ 4 × (i)	4			1	4	−1
16		1·164	0·010	0·124	1·24	7·60	3·04
17	+ (ii)		1		3	−2	1
18		1·164	0·011	0·124	3·26	−2·34	0·82

The residuals on line 2 are calculated from the rounded off operations table, but in the check on line 3 are calculated exactly from the original equations. *The residuals on line* 3 are then multiplied by 10 and rounded off again to the nearest integer on line 4.

The changes made in the values of the unknowns on lines 5 and 6 now affect the first decimal place as is shown on line 7. Although there has not yet been any change in the value of y it is written as 0·0 on this line so that there is no confusion about which decimal place has been reached. When the residuals are calculated exactly at this checking stage, by substituting $x = 1{\cdot}2$, $y = 0$ and $z = 0{\cdot}1$ in the original equations, the values obtained are $R_1 = 0{\cdot}330$, $R_2 = 0{\cdot}209$ and $R_3 = -\ 0{\cdot}288$. However, as the working is here in terms of the first decimal place, they are written on line 7 as 3·30, 2·09 and − 2·88.

These values are multiplied by 10 and rounded off to the nearest integer on line 8 and the process continued to line 12.

The true values of the residuals at this stage will be found to be − 0·0230, − 0·0198 and − 0·0534, but as the working is here in terms of the second decimal place they are written as − 2·30, − 1·98 and − 5·34 on line 12.

At the next checking stage on line 16 it is seen that the true values of the residuals are larger than expected, and can be reduced further using operation (ii) on line 17, causing a modification in the value of y.

The approximations to the values of the unknowns on line 18 agree to two decimal places with those on line 12.

Thus, the values of the unknowns, correct to 2D, are $x = 1{\cdot}16$, $y = 0{\cdot}01$ and $z = 0{\cdot}12$.

Any large difference in the values of the residuals at the checking stage may also be due to mistakes in the previous working. If so, it is usually better to continue the working with the values of the residuals given by the check, rather then to try to locate and correct such mistakes.

Examples for exercise are to be found in questions 1– 4, 10 in § 6.6.

6.4.2 Group operations

If a set of equations does not have a dominant leading diagonal then there will not be unit operations which can be used to reduce each residual whilst not having a great effect on the others. Frequently this difficulty can be overcome by constructing linear combinations of the unit operations, called *group operations* which will have this effect. Difficulties caused in ill-conditioned sets of equations however are not easily overcome.

Worked example 1.

Consider the following set of equations:

$$7x + 2y - 3z = 4$$
$$3x + 2y + z = 3$$
$$x - 3y + 6z = 5$$

Operations table:

No.	δx	δy	δz	δR_1	δR_2	δR_3	
(i)	1			7	3	1	$d_1 = \frac{4}{7}$
(ii)		1		2	2	−3	$d_2 = \frac{5}{2}$
(iii)			1	−3	1	6	$d_3 = \frac{2}{3}$

The unit operations (i) and (iii) are satisfactory for reducing the residuals R_1 and R_3 respectively, but operation (ii) is of little use in reducing R_2, in fact it has a greater effect on R_3 than on R_2. However, the combination of $2 \times$ (ii) + (iii) results in changes in the residuals of 1, 5 and 0 respectively, giving a good operation for reducing R_2, with $d_2 = \frac{1}{5}$. The complete operations table is then as follows:

No.	δx	δy	δz	δR_1	δR_2	δR_3	
(i)	1			7	3	1	$d_1 = \frac{4}{7}$
(ii)		1		2	2	-3	
(iii)			1	-3	1	6	$d_3 = \frac{2}{3}$
(iv)		2	1	1	5	0	$d_2 = \frac{1}{5}$

Lack of direct method

There is no direct method of finding such group operations. One or two may be easily obtained by inspection of the unit operations table, but it will probably be necessary to try several different combinations of the unit operations before one with the desired property is found. These can all be noted down in an extended operations table, as although only a few will be used in most of the working, others may be useful at certain stages.

The working for this set of equations is as follows:

Operation	x	y	z	R_1	R_2	R_3
	0	0	0	-4	-3	-5
+ (iii)			1	-7	-2	1
+ (i)	1			0	1	2
	1	0	1	√	√	√
				0	10	20
$- 3 \times$ (iii)			-3	9	7	2
− (i)	-1			2	4	1
− (iv)		-2	-1	1	-1	1
	0·9	$-0·2$	0·6	√	√	√
				10	-10	10
$+ 2 \times$ (iv)		4	2	12	0	10
$- 2 \times$ (i)	-2			-2	-6	-8
− (iii)			-1	1	-7	2
+ (iv)		2	1	2	-2	2
+ (ii)		1		4	0	-1
	0·88	$-0·13$	0·62	√	√	√
				40	0	-10
$- 5 \times$ (i)	-5			5	-15	-15
$+ 3 \times$ (iv)		6	3	8	0	-15
$+ 2 \times$ (iii)			2	2	2	-3
− (ii)		-1		0	0	0
	0·875	$-0·125$	0·625	√	√	√

The values of the unknowns have been obtained exactly since all the residuals have been reduced to zero.

Although operation (ii) does not affect one particular residual to the exclusion of the others it has still been used. The unit operations are rarely completely redundant, as illustrated again in the worked example 2.

Examples for exercise are to be found in questions 8 and 9 of § 6.6.

Worked example 2.

Find, correct to 2D, the values of x, y and z which satisfy the equations

$$23{\cdot}4x - 10{\cdot}5y - 52{\cdot}6z = 21{\cdot}5$$

$$36{\cdot}1x + 21{\cdot}8y - 32{\cdot}1z = 39{\cdot}7$$

$$52{\cdot}3x + 45{\cdot}7y + 28{\cdot}3z = 22{\cdot}9$$

These equations are first divided by 10, and for the operations table the coefficients are then rounded off to the nearest integer:

No.	δx	δy	δz	δR_1	δR_2	δR_3
(i)	1			2	4	5
(ii)		1		-1	2	5
(iii)			1	-5	-3	3

Operation (ii) can be used for reducing R_3, for which $d_3 = \frac{3}{5}$, but the other two unit operations are not very useful for reducing either R_1 or R_2.

Group operations for reducing these residuals must therefore be sought, but NOT on the basis of the above unit operations table. They must first be found accurately and then rounded off. For example, from the above table $-4 \times$ (i) $+ 5 \times$ (ii) $- 2 \times$ (iii) giving $\delta R_1 = 3$, $\delta R_2 = 0$ and $\delta R_3 = -1$ may seem a good operation for reducing R_1, but calculating the effects on the residuals gives $\delta R_1 = -4{\cdot}09$, $\delta R_2 = 2{\cdot}88$ and $\delta R_3 = -3{\cdot}73$, i.e. $\delta R_1 = -4$, $\delta R_2 = 3$ and $\delta R_3 = -4$ on rounding off. This operation thus has quite a different effect to that desired.

Two group operations which are useful here are:

firstly, operation (iv) $= -$ (i) $+ 2 \times$ (ii) $-$ (iii), giving $\delta R_1 = 0{\cdot}82$, $\delta R_2 = 3{\cdot}96$ and $\delta R_3 = 1{\cdot}08$, i.e. $\delta R_1 = 1$, $\delta R_2 = 4$ and $\delta R_3 = 1$ on rounding off, which can be used for reducing R_2, with $d_2 = \frac{1}{2}$;

secondly, operation (v) $= -4 \times$ (i) $+ 5 \times$ (ii) $-$ (iii), giving $\delta R_1 = -9{\cdot}35$, $\delta R_2 = -0{\cdot}33$ and $\delta R_3 = -0{\cdot}90$, i.e. $\delta R_1 = -9$, $\delta R_2 = 0$ and $\delta R_3 = -1$ on rounding off, which can be used for reducing R_1, with $d_1 = \frac{1}{9}$.

The complete operations table and working is thus as follows:

No.	δx	δy	δz	δR_1	δR_2	δR_3	
(i)	1			2	4	5	
(ii)		1		−1	2	5	$d_3 = \frac{3}{5}$
(iii)			1	−5	−3	3	
(iv)	−1	2	−1	1	4	1	$d_2 = \frac{1}{2}$
(v)	−4	5	−1	−9	0	−5	$d_1 = \frac{1}{9}$

Operation	x	y	z	R_1	R_2	R_3
	0	0	0	−2	−4	−2
+ (i)	1			0	0	3
	1	0	0	0·19	−0·36	2·94
				2	−4	29
− 6 × (ii)		−6		8	−16	−1
+ 4 × (iv)	−4	8	−4	12	0	3
+ (v)	−4	5	−1	3	0	2
	0·2	0·7	−0·5	2·13	−1·17	5·40
− (i)	−1			0	−5	0
	0·1	0·7	−0·5	0·21	−4·78	0·17
				2	−48	2
+ 12 × (iv)	−12	24	−12	14	0	14
+ 2 × (v)	−8	10	−2	−4	0	12
− 2 × (ii)		−2		−2	−4	2
+ (iv)	−1	2	−1	−1	0	3
	−0·11	1·04	−0·65	−8·04	−1·34	4·80
− (v)	4	−5	1	1	−1	5
− (ii)		−1		2	−3	0
+ (iv)	−1	2	−1	3	1	1
	−0·08	1·00	−0·65	3·18	0·77	2·21
				32	8	22
+ 3 × (v)	−12	15	−3	5	8	19
− 3 × (ii)		−3		8	2	4
+ (v)	−4	5	−1	−1	2	3
− (ii)		−1		0	0	−2
	−0·096	1·016	−0·654	−1·40	−2·34	0·22
				−14	−23	2
+ 5 × (iv)	−5	10	−5	−9	−3	7
− (v)	4	−5	1	0	−3	8
− 2 × (ii)		−2		2	−7	−2
+ 2 × (iv)	−2	4	−2	4	1	0
	−0·0963	1·0167	−0·6546	3·19	0·29	1·52

Notes on worked example 2.

Again the residuals are calculated exactly at each check line, it being necessary in two cases to reduce the residuals further before continuing to the next decimal place. Five iterations are required altogether here to obtain two successive approximations agreeing to two decimal places in all the unknowns, due to the rounding off errors in the operations table.

Thus $x = -0{\cdot}10$, $y = 1.02$ and $z = -0{\cdot}65$ correct to 2D.

Examples for exercise are to be found in questions 11– 15 of § 6.6.

6.5 THE GAUSS-SEIDEL METHOD

This is another iterative method, but its application is restricted to sets of equations which have a dominant leading diagonal.

A first approximation to the value of each unknown is found by including only the dominant and constant terms in each equation, and then each equation is rearranged so that the unknown with the dominant coefficient is the subject.

A second approximation to the value of the first unknown can then be found by substituting the first approximations to the values of the other unknowns in the first of these equations.

A better approximation for the second unknown is now found from the second equation, using the second approximation now available for the first unknown and the first approximations for the others.

Similar substitutions are made in the other equations, always using the latest approximation obtained for any unknown.

The process may then be repeated to obtain better approximations until two successive approximations agree to the accuracy required. To ensure this accuracy two extra figures should be retained in the working in the later stages.

Worked example.

Find, correct to 2D, the values of x, y and z which satisfy the equations

$$4x - y + z = 3 \qquad (1)$$

$$2x + 5y - z = 2 \qquad (2)$$

$$x + 2y + 6z = 5 \qquad (3)$$

First approximations. Ignoring the terms in y and z equation (1) gives $x_0 = 0{\cdot}8$. Similarly equation (2) gives $y_0 = 0{\cdot}4$ and equation (3) gives $z_0 = 0{\cdot}8$.

The rearranged equations give:

$$x_{r+1} = 0{\cdot}75 + 0{\cdot}25y_r - 0{\cdot}25z_r$$

$$y_{r+1} = 0{\cdot}4 - 0{\cdot}4x_{r+1} + 0{\cdot}2z_r$$

$$z_{r+1} = 0{\cdot}8\dot{3} - 0{\cdot}1\dot{6}x_{r+1} - 0{\cdot}\dot{3}y_{r+1}$$

So that it is easier to carry out the necessary working it is set out in tabular form, with each set of approximations in a column, as follows:

		1	2	3	4	5
1	0·75	0·8	0·75	0·75	0·75	0·75
2	$0{\cdot}25y_r$	0·1	0·08	0·063	0·0668	0·0660
3	$-0{\cdot}25z_r$	−0·2	−0·15	−0·16	−0·1588	−0·1589
4	x_{r+1}	0·7	0·68	0·653	0·6580	0·6571
5	0·4	0·4	0·4	0·4	0·4	0·4
6	$-0{\cdot}4x_{r+1}$	−0·3	−0·27	−0·261	−0·2632	−0·2628
7	$0{\cdot}2z_r$	0·2	0·12	0·128	0·1270	0·1271
8	y_{r+1}	0·3	0·25	0·267	0·2638	0·2643
9	0·8333	0·8	0·83	0·833	0·8333	0·8333
10	$-0{\cdot}1667x_{r+1}$	−0·1	−0·11	−0·109	−0·1097	−0·1095
11	$-0{\cdot}3333y_{r+1}$	−0·1	−0·08	−0·089	−0·0880	−0·0881
12	z_{r+1}	0·6	0·64	0·635	0·6356	0·6357

Thus, correct to 2D, $x = 0{\cdot}66$, $y = 0{\cdot}26$, $z = 0{\cdot}64$.

Notes on the operation table

The full number of decimal places required is not used in the first stages of the working, in order to simplify it.

In column 1 on lines 2 and 3 the first approximations to y and z, 0·4 and 0·8, are used to calculate $0{\cdot}25y_0$ and $-0{\cdot}25z_0$, giving the second approximation to x as 0·7, on line 4. This new value of x is used to calculate $-0{\cdot}4x_1$ on line 6, but only the first approximation to z can be used on line 7 to calculate $0{\cdot}2z_0$, thus giving the second approximation to y as 0·3, on line 8. These second approximations to x and y are used on lines 10 and 11 giving the second approximation to z on line 12.

Similarly in column 2 the second approximations to y and z are used on lines 2 and 3, but the third approximation to x is available for line 6.

The process is continued in the same way, always using the latest approximation to the value of any unknown, until in columns 4 and 5 the values of the unknowns are the same correct to 2D.

The values obtained should be checked by substitution in the *original* equations as it is possible to make a mistake in rearranging the equations which does not affect the convergence of the solution.

The Gauss-Seidel method should be compared with other iterative methods for the solution of equations given in § 4.5.

The description of these two iterative methods for the solution of linear simultaneous equations has here been limited to a discussion of the basic principles. Both methods are capable of refinements, and the inherent concepts are the basis of methods which are used in more advanced work.

6.6 EXAMPLES

Find the values of the unknowns in the following sets of equations to the given accuracy.

1. $8{\cdot}7x - 2{\cdot}3y = 4{\cdot}8$
   $3{\cdot}2x + 7{\cdot}4y = 5{\cdot}3$ (3D)

2. $7{\cdot}24x - 2{\cdot}03y = 6{\cdot}75$
   $3{\cdot}11x - 6{\cdot}18y = -3{\cdot}47$ (3D)

3. $4{\cdot}37x + 2{\cdot}87y = 5{\cdot}42$
   $2{\cdot}19x - 6{\cdot}91y = 3{\cdot}07$ (3D)

4. $4{\cdot}46x + 1{\cdot}25y = 3{\cdot}64$
   $2{\cdot}31x - 5{\cdot}18y = 2{\cdot}07$ (4D)

5. $5x + y - 2z = 5$
   $3x + 9y + 5z = 9$
   $x - y + 8z = 5$ (3D)

6. $5x + y + 3z = 3$
   $x - 4y + z = 4$
   $2x + y - 6z = 5$ (3D)

7. $8x + 3y - z = 2$
   $2x - 6y - 2z = 5$
   $x + y + 4z = 4$ (2D)

8. $2x + 4y - z = 8$
   $x - 3y + 2z = 7$
   $3x + 2y + 5z = 12$ (3D)

9. $6x - 2y + z = 7$
   $x + 5y - 2z = 6$
   $2x - y + z = 4$ (3D)

10. $8{\cdot}23x + 3{\cdot}67y - 2{\cdot}08z = 7{\cdot}34$
    $2{\cdot}55x - 5{\cdot}17y - 1{\cdot}62z = -1{\cdot}93$
    $3{\cdot}02x + 2{\cdot}18y + 7{\cdot}81z = 9{\cdot}23$ (2D)

11. $5{\cdot}2x - 3{\cdot}8y + 2{\cdot}2z = 14{\cdot}3$
$2{\cdot}9x + 2{\cdot}1y - 0{\cdot}8z = 3{\cdot}2$
$1{\cdot}9x + 3{\cdot}3y - 4{\cdot}1z = -\,9{\cdot}1$ (2D)

12. $7{\cdot}17x - 2{\cdot}53y - 1{\cdot}29z = 2{\cdot}62$
$1{\cdot}85x + 8{\cdot}78y + 1{\cdot}19z = -\,8{\cdot}74$
$3{\cdot}55x - 4{\cdot}16y + 5{\cdot}90z = -\,2{\cdot}59$ (3D)

13. $7{\cdot}32x + 1{\cdot}56y - 2{\cdot}04z = 14{\cdot}2$
$2{\cdot}16x + 9{\cdot}41y + 3{\cdot}49z = 21{\cdot}7$
$1{\cdot}48x - 4{\cdot}29y + 2{\cdot}46z = 16{\cdot}5$ (3D)

14. $9{\cdot}43x + 6{\cdot}84y + 3{\cdot}96z = -\,17{\cdot}2$
$4{\cdot}88x - 5{\cdot}47y + 2{\cdot}14z = 6{\cdot}8$
$2{\cdot}95x - 8{\cdot}44y + 9{\cdot}09z = -\,63{\cdot}1$ (3S)

15. $97x + 66y - 55z = 7{\cdot}94$
$45x - 33y - 23z = 8{\cdot}35$
$37x - 67y + 95z = 2{\cdot}02$ (3D)

16. Solve for x_1, x_2 and x_3 the following equations correct to 4D

$$11{\cdot}84x_1 + 9{\cdot}15x_2 + 2{\cdot}15x_3 = 6{\cdot}88$$
$$4{\cdot}26x_1 + 15{\cdot}36x_2 - 2{\cdot}89x_3 = -\,8{\cdot}61$$
$$6{\cdot}30x_1 - 5{\cdot}88x_2 + 3{\cdot}85x_3 = 12{\cdot}95$$

(*A.E.B.* 1960)

17. Solve for x_1, x_2, x_3 and x_4 correct to 4D

$$8x_1 + 3x_2 + x_3 = 2{\cdot}0307$$
$$3x_1 + 6x_2 - 5x_3 + x_4 = 0{\cdot}4811$$
$$x_1 - 5x_2 + 7x_3 + 2x_4 = 2{\cdot}1659$$
$$x_2 + 2x_3 + 4x_4 = 2{\cdot}5657$$

(*A.E.B.* 1962)

18. Solve for x_1, x_2 and x_3 correct to 2D

$$\frac{5{\cdot}22}{x_1} + \frac{1{\cdot}75}{x_2} - \frac{2{\cdot}13}{x_3} = 30{\cdot}47$$
$$\frac{3{\cdot}82}{x_1} + \frac{7{\cdot}11}{x_2} + \frac{4{\cdot}91}{x_3} = 34{\cdot}60$$
$$\frac{4{\cdot}76}{x_1} - \frac{0{\cdot}87}{x_2} + \frac{6{\cdot}41}{x_3} = 2{\cdot}40$$

(*A.E.B.* 1965)

7

Roots of Polynomial Equations

7.1 SYNTHETIC DIVISION

7.1.1 Division by a linear factor

Case 1

Coefficient of x in the factor is unity.

Worked example

Consider the division of $2x^3 + 4x^2 - 11x + 3$ by $x - 3$ by the usual method:

$$\begin{array}{rrl}
 & 2x^2 + 10x + 19 & \quad 1 \\
x - \mathbf{3}\,) & \mathbf{2}x^3 + \mathbf{4}x^2 - \mathbf{11}x + \mathbf{3} & \quad 2 \\
 & 2x^3 - \mathbf{6}x^2 & \quad 3 \\
 & \mathbf{10}x^2 - 11x & \quad 4 \\
 & 10x^2 - \mathbf{30}x & \quad 5 \\
 & \mathbf{19}x + 3 & \quad 6 \\
 & 19x - \mathbf{57} & \quad 7 \\
 & \mathbf{60} & \quad 8
\end{array}$$

which gives a quotient of $2x^2 + 10x + 19$ with remainder 60.

This method for the division of a polynomial contains several repetitions of coefficients and powers of x. The only essential part of the working consists of the numbers in bold type.

Note first of all how the coefficients required for the quotient, viz. 2, 10 and 19, appear in the working as the coefficients of x^3 in line 2, of x^2 in line 4 and of x in line 6. These are obtained from the coefficients of the polynomial as follows:

2 is the coefficient of x^3,

10 is $4 - (-3 \times 2)$,

19 is $-11 - (-3 \times 10)$

60, the remainder, is $3 - (-3 \times 19)$.

These can be evaluated easily using the following layout:

	2	4	−11	3	line 1
+3)		6	30	57	line 2
	2	10	19	60	line 3

Method: Write down the coefficients of the polynomial on the first line.

If dividing by $x - a$ write down $+a$ as shown on the second line, and the first coefficient on the third line.

Each entry in the second line is found by multiplying the previous entry in the third line by $+a$.

Each entry in the third line is found by adding the entries above it in the first two lines.

(Note that $+a$ is used as the multiplier so that the numbers in each column are always added instead of subtracted as in the original method above).

The flow chart for this method is as follows:

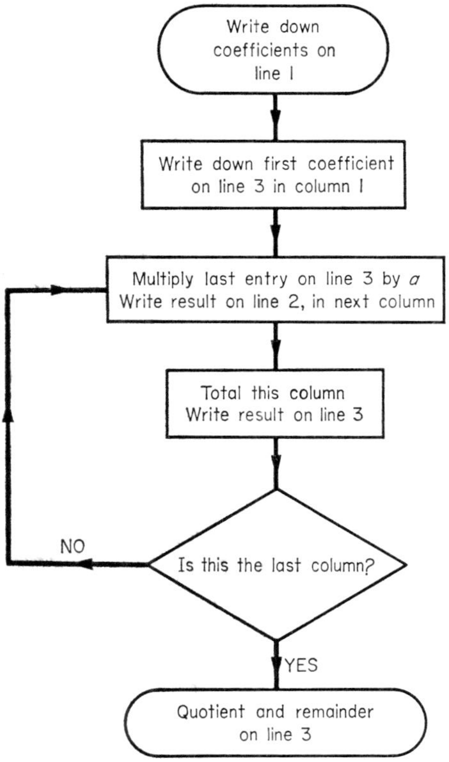

Division by a linear factor $x - a$ also gives the value of the polynomial at $x = a$, for when a polynomial of degree n is divided by $(x - a)$ then $f(x) = (x - a)q(x) + r$, where $q(x)$, a polynomial of degree $n - 1$, is the quotient and r, a constant, is the remainder. Putting $x = a$ it follows that $f(a) = r$. (This is known as the remainder theorem). Thus in the above example $f(3) = 60$.

In fact the sequence of values obtained in the synthetic division layout is the same as that used in the evaluation of a polynomial by nesting. Thus the successive terms in the evaluation of $2x^3 + 4x^2 - 11x + 3$ at $x = 3$ by nesting are $2 \times 3 + 4 = 10$, $10 \times 3 - 11 = 19$, $19 \times 3 + 3 = 60$.

If only a value of the polynomial is required then nesting is quicker as it does not require any recording of intermediate stages.

Worked example 2

$$3x^4 - x^3 + 2x^2 + x - 7 \div x + 2$$

	3	−1	2	1	−7
−2)		−6	14	−32	62
	3	−7	16	−31	55

which gives quotient $3x^3 - 7x^2 + 16x - 31$ remainder 55 and so $f(-2) = 55$.

Worked example 3

$$2x^3 + 1{\cdot}6x - 5{\cdot}3 \div x - 2$$

	2	0	1·6	− 5·3
2)		4	8	19·2
	2	4	9·6	13·9

which gives quotient $2x^2 + 4x + 9{\cdot}6$ remainder 13·9 and thus also $f(2) = 13{\cdot}9$. Note that *missing powers of x must be included.*

Case 2

Coefficient of x in the factor is not unity.

Consider the division of $f(x) = ax^n + bx^{n-1} + \ldots + k$ by $mx - n$, giving quotient $q(x)$ and remainder r.

Then $f(x) = (mx - n)q(x) + r$ and thus $f(x)/m = (x - [n/m])q(x) + r/m$, dividing by m. i.e. if $f(x)/m$ is divided by $(x - [n/m])$ the same quotient is obtained as for $f(x)$ divided by $mx - n$, but the remainder divided by m.

Note that the remainder r is equal to $f(n/m)$,

e.g. Compare $4x^2 + 3x - 1 \div 2x - 3$ and $2x^2 + 1{\cdot}5x - 0{\cdot}5 \div x - 1{\cdot}5$;

$$\begin{array}{r|l} & 2x + 4{\cdot}5 \\ \hline 2x - 3 & 4x^2 + 3x - 1 \\ & 4x^2 - 6x \\ \hline & \quad\;\; 9x - 1 \\ & \quad\;\; 9x - 13{\cdot}5 \\ \hline & \qquad\quad 12{\cdot}5 \end{array} \qquad \begin{array}{r|l} & 2x + 4{\cdot}5 \\ \hline x - 1{\cdot}5 & 2x^2 + 1{\cdot}5x - 0{\cdot}55 \\ & 2x^2 - \;\; 3x \\ \hline & \quad\; 4{\cdot}5x - 0{\cdot}5 \\ & \quad\; 4{\cdot}5x - 6{\cdot}75 \\ \hline & \qquad\quad 6{\cdot}25 \end{array}$$

i.e. The remainder in the second case must be doubled to give the correct result. *Conclusion* We can use the method of synthetic division for this type of example if the original terms are divided by, and the remainder multiplied by, the coefficient of x in the factor.

Worked example

For $4x^2 + 3x - 1 \div 2x - 3$ we have the following synthetic division layout:

$$\begin{array}{rccc} & 2 & 1{\cdot}5 & -0{\cdot}5 \\ 1{\cdot}5) & & 3 & 6{\cdot}75 \\ \hline & 2 & 4{\cdot}5 & 6{\cdot}25 \end{array}$$

so the quotient is $2x + 4{\cdot}5$ and the remainder $12{\cdot}5$ and so $f(1{\cdot}5) = 12{\cdot}5$.

Worked example 2

$$2x^4 - 4x^2 + 5x + 7 \div 5x + 2$$

Here the divisor and dividend are first divided by 5 so that the remainder must be multiplied by 5.

$$\begin{array}{rccccc} & 0{\cdot}4 & 0 & -0{\cdot}8 & 1 & 1{\cdot}4 \\ -0{\cdot}4) & & -0{\cdot}16 & 0{\cdot}064 & 0{\cdot}2944 & -0{\cdot}517\,76 \\ \hline & 0{\cdot}4 & -0{\cdot}16 & -0{\cdot}736 & 1{\cdot}2944 & 0{\cdot}882\,24 \end{array}$$

The quotient is $0{\cdot}4x^3 - 0{\cdot}16x^2 - 0{\cdot}736x + 1{\cdot}2944$ and the remainder is $5 \times 0{\cdot}882\,24 = 4{\cdot}4112$ so $f(-0{\cdot}4) = 4{\cdot}4112$.

Checking

A quick partial check of the result of a synthetic division can be made by substituting $x = 0$ in the original and final terms.

(*a*) *Check of Case* 1, *Example* 1

Consider again the division of $2x^3 + 4x^2 - 11x + 3$ by $x - 3$. The quotient is $2x^2 + 10x + 19$ and the remainder is 60, so

$$(2x^3 + 4x^2 - 11x + 3) \div (x - 3) = (2x^2 + 10x + 19) + 60/(x - 3)$$

Substituting $x = 0$ in the left hand side we have $3 \div -3 = -1$, and substituting $x = 0$ in the right hand side we have $19 + 60/-3 = -1$.

If the left hand and the right hand sides are not equal at $x = 0$ then there is some mistake in the working,

(*b*) *Check of Case* 2, *Example* 1

$4x^2 + 3x - 1 \div 2x - 3$ has quotient $2x + 4{\cdot}5$ and remainder $12{\cdot}5$.

Substituting $x = 0$ in the original terms we have $-1 \div -3 = \frac{1}{3}$. Substituting $x = 0$ in the quotient and remainder we have $4{\cdot}5 + 12{\cdot}5/-3 = \frac{1}{3}$.

Rounding off

If only limited accuracy is required the terms of the sequence may be rounded off, though it is desirable to carry two guarding figures in the working.

7.1.2 Examples

Use synthetic division in the following examples:

1. $x^3 + 2x^2 - 3x + 5 \div x - 2$
2. $x^4 - 6x^2 + 4x + 1 \div x + 3$
3. $3{\cdot}5x^2 + 2{\cdot}4x + 1{\cdot}3 \div x + 4$
4. $2x^3 - x^2 + 6x + 5 \div 2x - 5$
5. $4x^3 + 6x - 7 \div 3x + 6$
6. Show that $(x - 4)$ is a factor of $x^4 - 2x^3 - 9x^2 - x + 20$
7. Show that $(x - 2{\cdot}4)$ is a factor of $1{\cdot}7x^3 - 0{\cdot}58x^2 - 7{\cdot}1x - 3{\cdot}12$
8. Evaluate $3x^2 + 4x + 5$ at $x = 1{\cdot}6$ and $x = -2{\cdot}5$
9. Evaluate $x^3 - 3{\cdot}4x^2 + 2{\cdot}8x + 1{\cdot}7$ at $x = 1{\cdot}62$ to 4D
10. Evaluate $1{\cdot}68x^3 - 2{\cdot}89x + 0{\cdot}79$ at $x = 0{\cdot}6441$ to 3D
11. Write down the synthetic division scheme for $ax^3 + bx^2 + cx + d$ divided by $(x - x_0)$ and compare the succeeding stages of the solution with those obtained when evaluating a polynomial by the method of nested multiplication. (see §2.3).

7.1.3 Division by a quadratic factor

Worked example 1

Consider the division of $x^4 + 2x^3 + 4x^2 + 4x + 1$ by $x^2 + 3x + 5$. The coefficients essential to the working are printed in bold type.

$$
\begin{array}{r}
x^2 - x + 2 \\
x^2 + 3x + 5\,)\,\overline{\mathbf{1}x^4 + \mathbf{2}x^3 + \mathbf{4}x^2 + \mathbf{4}x + \mathbf{1}} \\
\underline{x^4 + \mathbf{3}x^3 + \mathbf{5}x^2} \\
-\,\mathbf{1}x^3 - x^2 + 4x \\
\underline{-\,x^3 - \mathbf{3}x^2 - \mathbf{5}x} \\
\mathbf{2}x^2 + 9x + 1 \\
\underline{2x^2 + \mathbf{6}x + \mathbf{10}} \\
\mathbf{3}x - \mathbf{9}
\end{array}
$$

This working can be reduced to the following scheme:

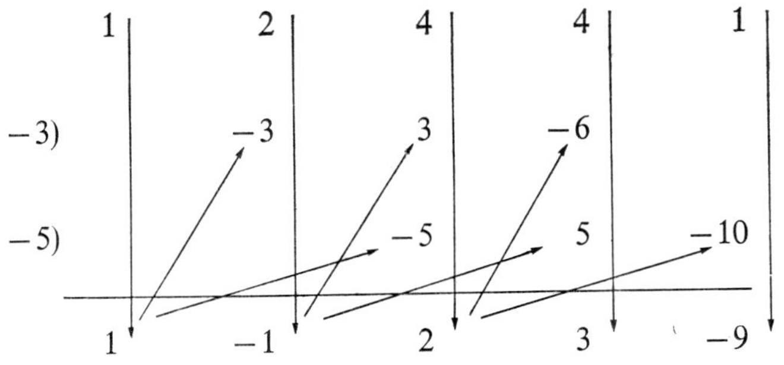

The coefficient of x and the constant term in the divisor are both written with their signs reversed in the second and third lines as shown, the change of sign in the coefficients again being made in order to avoid subtraction in the working.

The first column is totalled and the result, 1, multiplied first by -3, and this result written in the second column of the second line, second by -5, and this result written in the third column of the third line.

Repeat with the other columns but note that there should be no entry in the second line of the last column.

The remainder will usually consist of a term in x and a constant term and it is probably easier to pick out the quotient and remainder from the last line if the last two columns are marked off from the rest of the working by a vertical line as shown in the examples below.

The flow chart for the synthetic division of a polynomial by the quadratic factor $x^2 + bx + c$ is as follows:

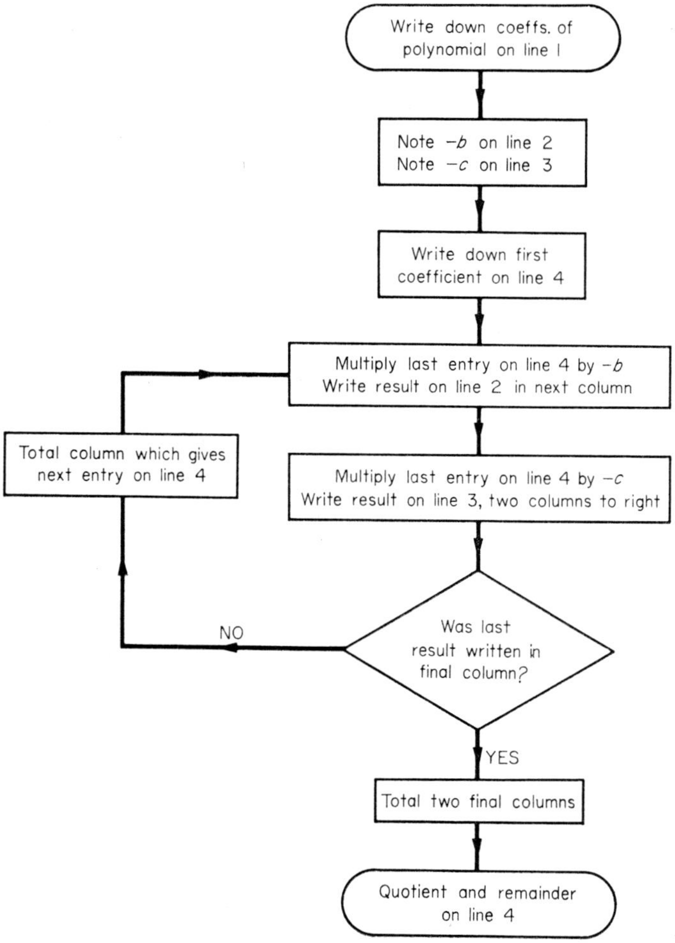

Worked example 2

$x^4 - 8x^3 + 5x^2 + 7x - 7 \div x^2 + x - 9$

Each stage of the working is shown below.

	1	−8	5	7	−7
−1)		−1			
9)			9		
	1				

	1	−8	5	7	−7
−1)		−1	9		
9)			9	−81	
	1	−9			

	1	−8	5	7	−7
−1)		−1	9	−23	
9)			9	−81	207
	1	−9	23		

	1	−8	5	7	−7
−1)		−1	9	−23	
9)			9	−81	207
	1	−9	23	−97	200

The quotient is $x^2 - 9x + 23$ and the remainder is $-97x + 200$.

Worked example 3

$$3x^3 + 5x^2 - 2x + 9 \div x^2 + 7$$

	3	5	− 2	9
0)		0	0	
− 7)			− 21	− 35
	3	5	− 23	− 26

The quotient is $3x + 5$ and the remainder is $-23x - 26$.

The second line could be omitted here but it is probably easier to follow the steps when it is included.

Worked example 4

$$4x^4 + 5x^3 + 3x^2 + 6x - 9 \div 2x^2 + 4x - 6$$

As for division by a linear factor the dividend and divisor are each divided by the coefficient of x^2 in the divisor. This again yields the correct quotient but the remainder must be multiplied by the original coefficient of x^2 in the divisor.

For this example $2x^4 + 2{\cdot}5x^3 + 1{\cdot}5x^2 + 3x - 4{\cdot}5$ is divided by $x^2 + 2x - 3$:

	2	2·5	1·5	3	− 4·5
− 2)		− 4	3	− 21	
3)			6	− 4·5	31·5
	2	− 1·5	10·5	− 22·5	27

so the answer to the original question is $2x^2 - 1{\cdot}5x + 10{\cdot}5$ remainder $-45x + 54$.

Worked example 5

$$2x^3 - 8x - 8 \div 5x^2 - 4x - 7$$

Here the coefficient of x^2 in the divisor is 5 so that both the divisor and dividend are first divided by 5, and the remainder finally multiplied by 5. *Note* that the missing term in x^2 in the polynomial must be included in the working.

	0·4	0	− 1·6	− 1·6
0·8)		0·32	0·256	
1·4)			0·56	0·448
	0·4	0·32	− 0·784	− 1·152

The quotient is $0{\cdot}4x + 0{\cdot}32$ and the remainder $-3{\cdot}92x - 5{\cdot}76$

7.1.4 Examples

1. $2x^4 + 2x^3 + 3x^2 + x + 8 \div x^2 + 3x + 2$
2. $3x^4 - x^3 + 7x^2 + 8 \div x^2 - 2x + 3$
3. $2x^4 + 6x^3 - 8x^2 - x + 7 \div x^2 - 3$
4. $8x^3 + 5x^2 + 4x - 5 \div x^2 + 4x - 5$
5. $2x^4 - 6x^3 + 8x^2 - x - 7 \div 4x^2 + 2x - 6$
6. $x^5 - 5x^4 + 2x^3 + x^2 - 3x + 7 \div x^2 - 3x + 2$

7. $4x^5 - 7x^4 + 2x^3 + 2x^2 - 3x - 6 \div 2x^2 + 3x - 4$
8. Show that $2x^2 - 2x + 1$ is a factor of $4x^4 - x^3 + 0{\cdot}5x + 0{\cdot}5$
9. Show that $1{\cdot}5x^2 + x + 1$ is a factor of $3x^4 + 6{\cdot}5x^3 + 3{\cdot}5x^2 + 2x - 1$
10. Show that $x^2 + x + 4$ is a factor of $x^5 - x^4 + 5x^3 - x^2 + 16x + 16$
11. Write down the synthetic division scheme for $px^4 + qx^3 + rx^2 + sx + t$ divided by $x^2 + bx + c$

7.1.5 Evaluation of a polynomial at a complex value*

To evaluate a polynomial $f(x)$ at $x = a + ib$ or at $x = a - ib$ it is divided by

$$[x - (a + ib)]\,[x - (a - ib)] = x^2 - 2ax + a^2 + b^2$$

and $x = a + ib$ or $a - ib$ is substituted in the remainder.

Justification: If on dividing $f(x)$ by $x^2 - 2ax + a^2 + b^2$ there is a quotient $q(x)$ and remainder $rx + s$, where r and s are constants, then

$$f(x) = (x^2 - 2ax + a^2 + b^2)q(x) + rx + s$$

hence $f(a + ib) = r(a + ib) + s$

or $f(a - ib) = r(a - ib) + s$

since the quadratic is zero for $x = a + ib$ and $x = a - ib$.

Note that $f(a + ib)$ is the complex conjugate of $f(a - ib)$.

Worked example 1

Evaluate $x^4 - 8x^3 + 5x^2 + 7x - 7$ at $x = 2 + i3$. The polynomial must be divided by $[x - (2 + i3)]\,[x - (2 - i3)] = x^2 - 4x + 13$ which is carried out as follows:

	1	− 8	5	7	− 7
4)		4	− 16	− 96	
− 13)			− 13	52	− 312
	1	− 4	− 24	− 37	− 319

The remainder is $-37x - 319$

Hence $f(2 + i3) = -37(2 + i3) - 319$

$= -74 - i111 - 319$

i.e. $f(2 + i3) = -393 - i111$

and $f(2 - i3) = -393 + i111$

* A knowledge of the properties of complex numbers is assumed in this section.

Worked example 2

Evaluate $3x^3 + 5x^2 - 2x + 9$ at $x = 0{\cdot}5 - i(\sqrt{2})$ to 4D. The polynomial is divided by $[x - (0{\cdot}5 - i\sqrt{2})][x - (0{\cdot}5 + i\sqrt{2})] = x^2 - x + 2{\cdot}25$.

	3	5	-2	9
1)		3	8	
$-2{\cdot}25$)			$-6{\cdot}75$	-18
	3	8	$-0{\cdot}75$	-9

The remainder is $-0{\cdot}75x - 9$

$$\text{Hence } f[0{\cdot}5 - i\sqrt{(2)}] = -0{\cdot}75[0{\cdot}5 - i\sqrt{(2)}] - 9$$

$$= -0{\cdot}375 + i0{\cdot}75\sqrt{(2)} - 9$$

$$\text{i.e. } f[0{\cdot}5 - i\sqrt{(2)}] = -9{\cdot}375 + i1{\cdot}0607 \text{ to 4D.}$$

7.1.6 Examples

1. Evaluate $2x^3 - x^2 + 5x + 1$ at $x = 2 + i\sqrt{(3)}$
2. Evaluate $x^4 + 2x^2 + 5x + 8$ at $x = 4 + i\sqrt{(2)}$
3. Show that $x - 1 - i3$ is a factor of $x^3 - 5x^2 + 16x - 30$
4. Show that $x - 1{\cdot}5 + i\sqrt{(3)}$ is a factor of $x^3 - 7x^2 + 17{\cdot}25x - 21$
5. Show that $x - 2 - i\sqrt{(2)}$ is a factor of $2x^4 - 11x^3 + 25x^2 - 22x + 6$
6. Evaluate $x^3 + 2x^2 - x + 3$ at $x = 1{\cdot}643 + i2{\cdot}238$ to 3D
7. Evaluate $x^3 + x^2 - 8x - 13$ at $x = 2{\cdot}02 + i0{\cdot}445$ to 3D

7.2 ROOTS OF POLYNOMIAL EQUATIONS: GENERAL POINTS

General iterative methods for the solution of equations, all applicable to the solution of polynomial equations, have already been considered. We now consider the solution of polynomial equations in more detail because of the importance of these equations.

7.2.1 Number and types of roots

A value of x which satisfies the polynomial equation $ax^n + bx^{n-1} + \ldots + k = 0$ is called a root of the equation. If the polynomial has roots $r_1, r_2 \ldots r_n$ it can be written in the form $(x - r_1)(x - r_2) \ldots (x - r_n)$. For example $x^2 - 3x + 2$ can be written as $(x - 1)(x - 2)$ and hence the polynomial equation $x^2 - 3x + 2 = 0$ has roots $x = 1$ and $x = 2$. A polynomial of degree n must have n roots, *real* or *complex* (see §3.1.1), but if the graph of the polynomial is drawn it does not follow that it must meet the X axis in n points.

Consider for example, (*a*) the sketch graph in Fig. 7.2.1(a) of $y = x^3 - 2x + 4$ This meets the x axis in only one point at $x = -2$ since

$$x^3 - 2x + 4 = (x + 2)(x^2 + 2x + 2)$$
$$= (x + 2)\{x - (1 + i)\}\{x - (1 - i)\}$$

i.e. This equation has only *one real* root, the other *two* roots being *complex*.

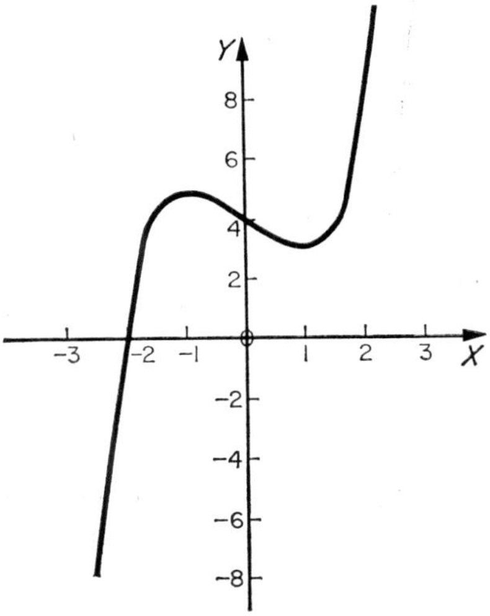

Fig. 7.2.1(a)

If the coefficients of a polynomial equation are real then the product of the factors $(x - r_1)(x - r_2) \ldots (x - r_n)$ must be real. Hence, if there are any complex roots, they must occur in conjugate pairs.

A polynomial equation of degree n will also have fewer than n distinct real roots if the x axis is a tangent to the curve. It will still be possible however to express the polynomial in the form $(x - r_1)(x - r_2) \ldots (x - r_n)$ where one or more pairs of roots are equal. Such roots are called *repeated roots*.

(*b*) The sketch graph in Fig. 7.2.1(b) of $x^2 - 4x + 4 = (x - 2)(x - 2) = 0$ shows it has *one pair* of *real repeated* roots.

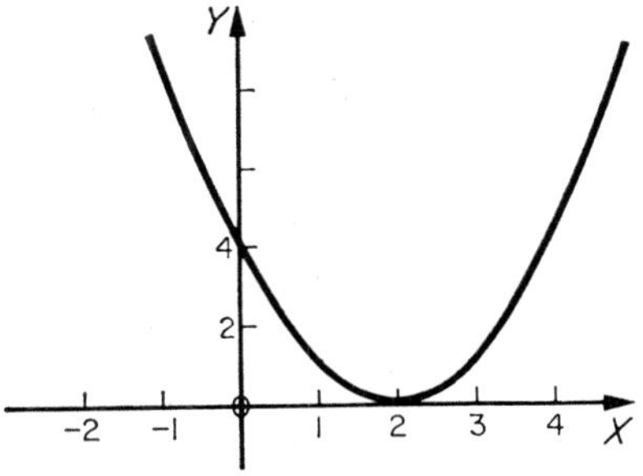

Fig. 7.2.1(b)

(*c*) The sketch graph in Fig. 7.2.1(c) of

$x^4 - 8x^2 + 16 = (x - 2)(x - 2)(x + 2)(x + 2) = 0$ shows it has *two pairs* of *real repeated* roots.

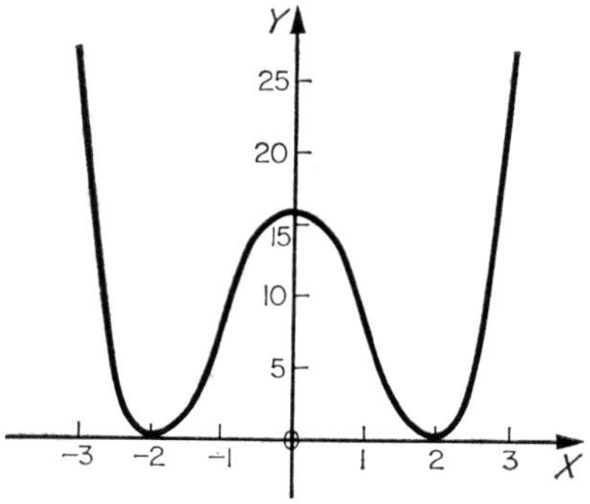

Fig. 7.2.1(c)

It is also possible for a polynomial equation to have *repeated complex* roots, but since these must occur in pairs if the polynomial has real coefficients, they can only occur in quartic equations and those of higher degree. The X axis is not a tangent to the curve in such cases.

If all the roots of a polynomial are required it is thus important to know of what *type* they may be. For example, a quadratic equation with real coefficients may have a pair of different real roots, a pair of repeated real roots or a conjugate pair of complex roots. A cubic equation with real coefficients will always have at least one real root, the other pair of roots having the same combinations as for a quadratic equation.

7.2.2 Approximate location of roots

Some idea of the types of root which an equation may have can be obtained from *Descartes' rule of signs*. This states that for an equation $f(x) = 0$ which has real coefficients there are no more positive roots than the number of changes in sign of the coefficients of $f(x)$, and no more negative roots than the number of changes in sign of the coefficients of $f(-x)$. The equation should be written in order of descending powers of x and missing terms ignored. For example $f(x) = x^3 - 5x^2 - 6x + 9 = 0$ has no more than two positive roots since $f(x)$ has two changes in sign, and no more than one negative root since $f(-x) = -x^3 - 5x^2 + 6x + 9$ has only one change in sign.

More accurate information is given by a neat sketch graph and it is usually best to begin the solution of a polynomial equation of degree three or higher with one, Descartes' rule of signs giving guidance as to the portion of the curve to sketch. As well as showing the nature of the roots such a sketch provides an initial approximation to the values of any real roots.

7.2.3 Attainable accuracy

For polynomial equations of low degree with exact coefficients the accuracy to which it is possible to obtain the roots is limited by the computing equipment available. However, as the order of the polynomial increases, or if any of the coefficients are rounded off, then the attainable accuracy is restricted. The discussion of methods of estimating the error in a value obtained for a root in such cases is beyond the scope of this book and the reader is referred to the bibliography, especially Redish, *Introduction to Computational Methods.*

7.3 ROOTS OF QUADRATIC EQUATIONS

The roots of the quadratic equation $ax^2 + bx + c = 0$ can usually be calculated using the formula $x = [-b \pm \sqrt{(b^2 - 4ac)}]/2a$. This shows that the quadratic equation will have different real roots when $b^2 > 4ac$, repeated real roots when $b^2 = 4ac$ and complex roots when $b^2 < 4ac$.

If a greater accuracy than four or five significant figures is required in the roots then it will be necessary to calculate the square root, rather than reading it from tables. If b^2 is much greater than $4ac$ the square root will have to be evaluated very accurately in order to obtain sufficient figures in the numerically smaller root. E.g. for the equation $x^2 + 10x + 0{\cdot}01 = 0$ $b^2 - 4ac = 100 - 0{\cdot}04 = 99{\cdot}96$ so that $\sqrt{(b^2 - 4ac)} = 9{\cdot}998$ correct to four significant figures, giving the numerically smaller root to be $\frac{1}{2}(-10 + 9{\cdot}998) = -0{\cdot}001$ which is only correct to one significant figure. This difficulty can be overcome by finding the numerically larger root by this method and then using Newton's method (see §4.6) to improve the value of the numerically smaller root. Alternatively the following iterative method can be used.

Let the roots be α and β. Then $\alpha + \beta = -b/a$ and $\alpha . \beta = c/a$. Since $b^2 \gg 4ac$, $\sqrt{(b^2 - 4ac)} \simeq b$ and it follows that one root (α say) must be small, and the other root (β) near to $-b/a$. Using $\beta = -b/a$ as a first approximation better values of the roots are obtained by substituting in the two equations $\alpha = (c/a) . (1/\beta)$ and $\beta = -(b/a) - \alpha$ in turn.

Worked example

Find the roots of $x^2 + 10x + 0{\cdot}02 = 0$ correct to 5S.

$$\alpha . \beta = 0{\cdot}02 \text{ thus } \alpha = \frac{0{\cdot}02}{\beta} \tag{1}$$

$$\text{and } \alpha + \beta = -10 \text{ thus } \beta = -10 - \alpha \tag{2}$$

The first approximation for β, $\beta_0 = -10$ ($-(b/a)$ as above) and the necessary working can be shown in tabular form as follows:

$\beta = -10 - \alpha$	$\alpha = 0{\cdot}02/\beta$
− 10	− 0·002
− 9·998	− 0·002 000 4
− 9·997 999 6	− 0·002 000 4

Thus correct to 5S $\alpha = -0{\cdot}002\,000\,4$ and $\beta = -9{\cdot}9980$. The roots obtained for a quadratic equation should always be checked to show that the sum of the roots $= -b/a$.

7.3.1 Examples

Find both roots of each of the following equations correct to 6S using the most appropriate method.

1. $x^2 + 10x - 1 = 0$
2. $x^2 + 9x + 21 = 0$
3. $x^2 + 16x - 0{\cdot}002 = 0$
4. $x^2 - 12x + 0{\cdot}01 = 0$
5. $x^2 - 4x + 10 = 0$
6. $x^2 - 7{\cdot}6x + 14{\cdot}44 = 0$
7. Assuming the numerical coefficients in the following equation to be exact, find each of the roots correct to 5S: $x^2 - 10x + 0{\cdot}005 = 0$. (*A.E.B.* 1960)

7.4 ROOTS OF CUBIC EQUATIONS

All cubic equations with real coefficients have at least one real root. This root (r_1 say) should be found using Newton's method (see §4.6), having obtained an initial approximation from a sketch graph. The cubic should then be divided by $x - r_1$ (see §7.1.1) to obtain a quadratic equation, the roots of which can be evaluated as in §7.3.

The complete solution of a cubic equation usually falls into three stages.
Stage one: sketching the curve to determine an initial approximation.
Stage two: calculation of one real root by Newton's method.
Stage three: calculation of the remaining pair of roots from the quadratic factor.

Worked example 1

To find all the roots of the equation: $x^3 - 5x^2 + 3x + 8 = 0$ correct to 3D.

Stage 1: Sketch graph.

Descartes' rule of signs (see §7.2.2) shows that there are no more than two positive roots and one negative root. As $x \to \infty$, $y \to \infty$ and as $x \to -\infty$, $y \to -\infty$.

Beginning with $x = 0$ values of the polynomial are calculated until either all the roots have been located or it is clear that $y \to \pm\infty$.

x	-2	-1	0	1	2	3	4
$f(x)$	-26	-1	8	7	2	-1	4

The sketch graph in Fig. 7.4(a) shows that there are roots near to $-0{\cdot}9$, $2{\cdot}5$ and $3{\cdot}5$.

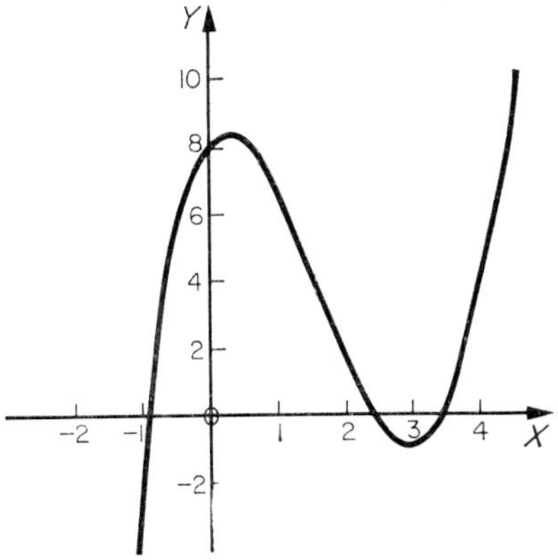

Fig. 7.4(a)

Stage 2: First root.

It is best here to find the root near to $-0{\cdot}9$ first as there is no other root near to it

Since $f(x) \equiv x^3 - 5x^2 + 3x + 8$ it follows that

$f'(x) \equiv 3x^2 - 10x + 3$ and thus from the formula (see §4.6) for Newton's method: $x_{r+1} = x_r - f(x_r)/f'(x_r)$ we have here

$$x_{r+1} = x_r - \frac{(x_r^3 - 5x_r^2 + 3x_r + 8)}{3x_r^2 - 10x_r + 3}.$$

The values of the numerator and denominator of the second term are calculated using nesting and the necessary working set out in tabular form as follows. Two extra guarding figures should be used in the working because of the number of operations necessary in the complete solution, so that five decimal places are retained in this example.

x_r	$f(x_r)$	$f'(x_r)$	$f(x_r) \div f'(x_r)$	x_{r+1}
$-0{\cdot}9$	$0{\cdot}521$	$14{\cdot}43$	$0{\cdot}036$	$-0{\cdot}936$
$-0{\cdot}94$	$-0{\cdot}068\,58$	$15{\cdot}050\,8$	$-0{\cdot}004\,56$	$-0{\cdot}935\,44$
$-0{\cdot}935\,4$	$0{\cdot}000\,485$	$14{\cdot}978\,92$	$0{\cdot}000\,03$	$-0{\cdot}935\,43$
$-0{\cdot}935\,43$	$0{\cdot}000\,04$	$14{\cdot}979\,39$	$0{\cdot}000\,00(3)$	$-0{\cdot}935\,43$

The first value in the x_{r+1} column shows that the initial approximation was correct to one significant figure, and so the second approximation is taken to two significant figures. Since it is likely that it will be possible to obtain the third approximation correct to four significant figures five decimal places are retained in the second line. The value obtained for x_{r+1} shows that the second approximation was correct to two significant figures so that four are retained in the third approximation. The working is completed using five decimal places, the last line serving as a check. The value $-0{\cdot}935\,43$ for the first root is thus used in calculating the other two roots.

Alternative working.

The expression obtained by substitution in the formula for Newton's method may be simplified as follows:

$$x_{r+1} = x_r - \frac{(x_r^3 - 5x_r^2 + 3x_r + 8)}{3x_r^2 - 10x_r + 3}$$

$$= \frac{3x_r^3 - 10x_r^2 + 3x_r - x_r^3 + 5x_r^2 - 3x_r - 8}{3x_r^2 - 10x_r + 3}$$

hence $$x_{r+1} = \frac{2x_r^3 - 5x_r^2 - 8}{3x_r^2 - 10x_r + 3}$$

and the working is then:

x_r	$2x_r^3 - 5x_r^2 - 8$	$3x_r^2 - 10x_r + 3$	x_{r+1}
−0·9	−13·508	14·43	−0·936
−0·94	−14·079 17	15·050 8	−0·935 44
−0·935 4	−14·011 77	14·978 92	−0·935 43
−0·935 43	−14·012 20	14·979 39	−0·935 43

The number of figures retained in the working at each stage is exactly the same as for the first method.

The choice of method is a matter of personal preference. The first has the advantage that as better approximations to the value of the root are obtained the value of the polynomial can be seen to be approaching zero. On the other hand the second method is quicker, as it requires one less operation at each iteration.

Stage 3: Remaining roots.

The quadratic factor is first found by synthetic division as follows. Note that this checks the value of the first root by showing that the remainder is negligible.

	1	−5	3	8
−0·935 43)		−0·935 43	5·552 18	−7·999 96
	1	−5·935 43	8·552 18	0·000 04

Thus the quadratic factor is $x^2 - 5{\cdot}935\,43x + 8{\cdot}552\,18$ which has the roots

$$x = \tfrac{1}{2}[5{\cdot}935\,43 \pm \sqrt{(35{\cdot}229\,33 - 34{\cdot}208\,72)}]$$
$$= \tfrac{1}{2}[5{\cdot}935\,43 \pm \sqrt{(1{\cdot}020\,61)}]$$
$$= \tfrac{1}{2}(5{\cdot}935\,43 \pm 1{\cdot}010\,25)$$
$$= \tfrac{1}{2}(6{\cdot}945\,68 \text{ or } 4{\cdot}925\,18)$$

hence $x = 3{\cdot}472\,84$ or $2{\cdot}462\,59$

And so the roots are $-0{\cdot}935$, $2{\cdot}463$ and $3{\cdot}473$ correct to 3D. If all the roots of a cubic equation are real, their values can be checked by noting that if the coefficient of x^3 is a, and the constant term is c, then the product of the roots should be $-c/a$, or they can be checked by substitution in the original equation.

Worked example 2

Find all the roots of the equation

$$x^3 + 3x^2 - 2x - 12 = 0 \text{ correct to 3D.}$$

Stage 1: Sketch graph.

Descartes' rule of signs shows that there are no more than one positive and two negative roots. The table of values of the cubic shows no negative roots in the region $x = 0$ to $x = -4$ but as it is clear that $f(x)$ is getting large and negative, and as $f(x) \to -\infty$ when $x \to -\infty$ no further values are calculated.

x	-4	-3	-2	-1	0	1	2	3
$f(x)$	-20	-6	-4	-8	-12	-10	4	36

The sketch graph in Fig. 7.4(b) shows that there is a real root near to 1·8 and that the other two roots are complex.

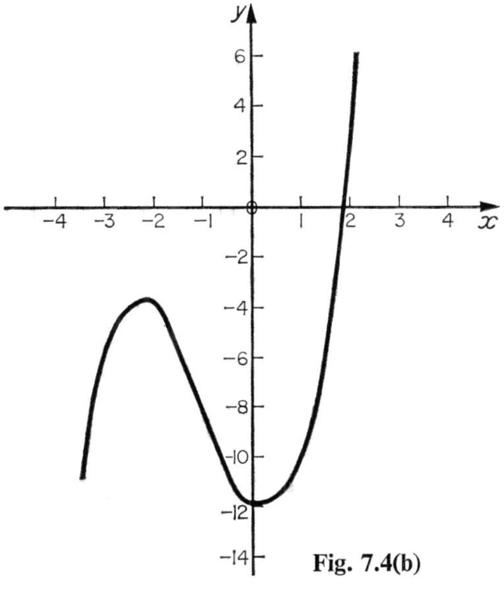

Fig. 7.4(b)

Stage 2: First root

Since $f(x) = x^3 + 3x^2 - 2x - 12$ it follows that $f'(x) = 3x^2 + 6x - 2$ and from the formula for Newton's method we have:

$$x_{r+1} = x_r - \frac{(x_r^3 + 3x_r^2 - 2x_r - 12)}{3x_r^2 + 6x_r - 2}$$

$$= \frac{3x_r^3 + 6x_r^2 - 2x_r - x_r^3 - 3x_r^2 + 2x_r + 12}{3x_r^2 + 6x_r - 2}$$

hence $$x_{r+1} = \frac{2x_r^3 + 3x_r^2 + 12}{3x_r^2 + 6x_r - 2}$$

Using the initial approximation 1·8 the real root is calculated as follows, retaining five decimal places in the working.

x_r	$2x_r^3 + 3x_r^2 + 12$	$3x_r^2 + 6x_r - 2$	x_{r+1}
1·8	33·384	18·52	1·802 59
1·802 59	33·462 41	18·563 53	1·802 59

Since the first value obtained for x_{r+1} shows that the first approximation was correct to three significant figures, six significant figures (here five decimal places) are retained for the second approximation. These six figures are confirmed in the last line of the working so that the value 1·802 59 for the first root is used to calculate the other two roots.

Stage 3: Remaining roots.

The quadratic factor is found using synthetic division, also checking the first root.

	1	3	−2	−12
1·802 59)		1·802 59	8·657 10	12·000 02
	1	4·802 59	6·657 10	0·000 02

The quadratic factor is $x^2 + 4{\cdot}802\,59x + 6{\cdot}657\,10$ and the two remaining roots are given by:

$$x = \tfrac{1}{2}\,[-4{\cdot}802\,59 \pm \sqrt{(23{\cdot}064\,87 - 26{\cdot}628\,40)}]$$

$$= \tfrac{1}{2}\,[-4{\cdot}802\,59 \pm \mathrm{i}\sqrt{(3{\cdot}563\,53)}]$$

$$= \tfrac{1}{2}\,(-4{\cdot}802\,59 \pm \mathrm{i}\;1{\cdot}887\,73)$$

hence $$x = -2{\cdot}401\,30 \pm \mathrm{i}\;0{\cdot}943\,87$$

The roots are 1·803 and $-2{\cdot}401 \pm \mathrm{i}\;0{\cdot}944$ correct to 3D.

The values obtained for such complex roots can be checked by noting that if the real root is r and the complex roots are $p \pm iq$ then $r\,(p^2 + q^2) = -(c/a)$. Where a is the coefficient of x^3 and c the constant term in the original cubic.

7.4.1 Examples

Find all the roots of the following equations:

1. $x^3 - x^2 - 2x + 1 = 0$ to 3D
2. $x^3 - 4x^2 + 3x + 2 = 0$ to 6D
3. $2x^3 - 5x^2 - 2x + 2 = 0$ to 3D
4. $x^3 - 2x^2 - x + 4 = 0$ to 4D
5. $x^3 + 3x^2 - x - 7 = 0$ to 3D
6. $x^3 - 15x + 10\sqrt{(5)} = 0$ to 5D

7.5 POLYNOMIAL EQUATIONS OF HIGHER DEGREE

If a quartic equation has a pair of real roots these can be found using Newton's method. Division of the quartic by the factors obtained gives a quadratic equation, the roots of which are found from the usual formula. Other methods of solution, which also apply to quartic equations having no real roots, are given in several of the books in the bibliography—see especially the method of resolving a quartic into two quadratic factors given in Butler and Kerr, *An Introduction to Numerical Methods.*

Quintic equations with real coefficients always have at least one real root, and this can be found using Newton's method. Division by the linear factor obtained from this root reduces the equation to a quartic.

Much labour is involved in solving equations of a higher degree using only a desk calculating machine but some of the books in the bibliography give methods which can be used if necessary.

7.5.1 Examples

Division by the quadratic factors given in the first two questions (see 7.1.3) reduces the first equation to a quadratic and the second to a cubic .

1. $z^4 + 1{\cdot}12z^3 - 11{\cdot}9364z^2 + 58{\cdot}52z - 69{\cdot}8896 = 0$ is a quartic with two real roots.

   $z^2 - 2{\cdot}94z + 8{\cdot}36$ is a factor of the left hand side. Find the other quadratic factor and hence evaluate the four roots correct to three significant figures. (The coefficients in the quartic are exact). (*A.E.B.* 1959)

2. Show that $x^2 - 9{\cdot}8x + 18{\cdot}25$ is a factor of

   $f(x) = x^5 - 8{\cdot}8x^4 + 13{\cdot}45x^3 - 25{\cdot}75x^2 + 42{\cdot}25x + 91{\cdot}25$.

   Hence find the three real roots and two complex roots of $f(x) = 0$. Give all answers correct to 2D. (The coefficients of $f(x)$ are all exact). (*A.E.B.* 1961)

3. Find all the roots of $4x^4 + 14x^3 + 14x^2 + 13x - 3 = 0$ correct to 3D.

8
Linear Interpolation

8.1 THE USE OF PROPORTIONAL PARTS

In this chapter we consider the simplest method of calculating values of a function between two given function values. This method is known as linear interpolation. In mathematical tables of elementary functions such as logarithms, trigonometrical tables and Napierian logarithms, when we use the mean differences we perform a kind of linear interpolation.

For example consider the following extract from 5-figure tables:

$\tan 50° 30'$	given as	1·213 10
$\tan 50° 36'$	given as	1·217 42

Here the difference between the two values is 0·004 32, which is usually written as 432 (see Chap. 5). Suppose we now require the value of the tangent of an angle between the two values given above. We may split the difference into six proportional parts giving 72, 144, 216, 288, 360, 432 respectively and by using these values and being aware that tangent is an increasing function in this range, we have

$$\begin{aligned}\tan 50° 31' &= \tan 50° 30' + \text{proportional part for 1 minute}\\ &= 1{\cdot}213\,10 \quad + 0{\cdot}000\,72\\ &= 1{\cdot}213\,82\end{aligned}$$

etc.

Thus we have performed linear interpolation using proportional parts. Most elementary tables are arranged for this method, although using mean differences, as explained in the following section.

8.1.1 Mean and true differences

Looking at the 5-figure tables for tan 50° we see that the mean differences given are 72, 143, 216, 288, 359 which differ from the proportional parts we obtained between 50° 30′ and 50° 36′ in §8.1. These discrepancies occur because the *mean* differences are evaluated from the average or mean of all the differences along the 50° row, which is the mean of the nine first differences

424, 425, 426, 428, 431, 432, 434, 436, 438. This mean is 430·55 and is split into the six proportional parts which are used.

Thus using true differences we have

$$\tan 50° \, 32' = 1{\cdot}214\,54$$

but using mean differences

$\tan 50° \, 32' = 1{\cdot}214\,53$ with discrepancy of 1 unit in the last significant digit.

Hence if full accuracy is required with linear interpolation true differences between adjacent tabular entries should be used.

The graphical theory of this method is based on the assumption that the curve between the two consecutive tabular points approximates to a straight line. This is considered further in the next section.

8.2 FORMULA FOR LINEAR INTERPOLATION

Using the notation of Chapter 5 we consider two adjacent entries in a table of differences for a general function, with values (x_0, f_0); (x_1, f_1). Thus these are the coordinates of two adjacent points on a curve passing through the tabulated function values. Let these points be $A(x_0, f_0)$; $B(x_1, f_1)$.

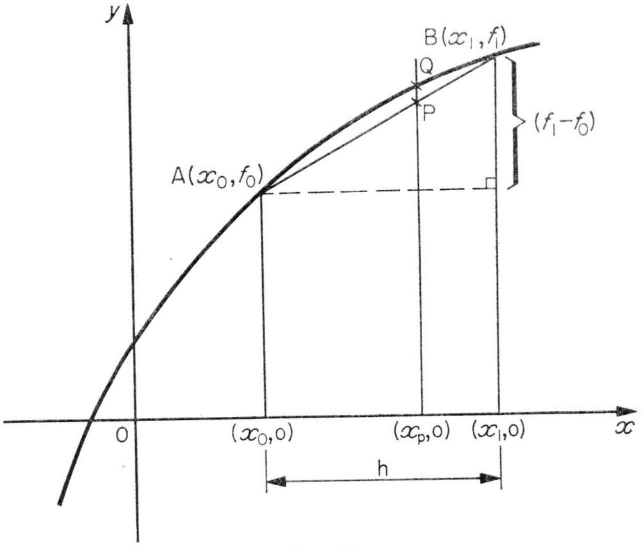

Fig. 8.2

We require to estimate the value of the function when the value of the independent variable is x_p. The true answer would be the y-coordinate of the point Q shown in the diagram (Fig. 8.2), but in linear interpolation we find the y co-ordinate of the point P on the straight line AB.

The gradient of AB $= \dfrac{f_1 - f_0}{x_1 - x_0}$

where $x_1 - x_0 = h$ is the tabular interval.

Therefore the gradient of AB $= \dfrac{f_1 - f_0}{h}$

giving the equation of AB as

$$y - f_0 = \frac{f_1 - f_0}{h}(x - x_0)$$

or

$$y = f_0 + \frac{f_1 - f_0}{h}(x - x_0)$$

but $f_1 - f_0$ may be written as Δf_0, the first order difference, giving

$$y = f_0 + \frac{\Delta f_0}{h}(x - x_0)$$

Using this equation to find the value of the function at P, when $x = x_p$ gives

$$y_p = f_0 + \frac{\Delta f_0}{h}(x_p - x_0)$$

Further, the value x_p can be represented in the form

$$x_p = x_0 + p.h \tag{1}$$

where 'p' will always lie in the range $0 < p < 1$ and is known as the *interpolating factor*.

Substituting this value for x_p in the equation for y_p gives

$$y_p = f_0 + \frac{\Delta f_0}{h}(x_0 + p.h - x_0)$$

$\therefore \quad y_p = f_0 + p.\Delta f_0$ and to keep the notation consistent we have,

$$f_p = f_0 + p.\Delta f_0 \tag{2}$$

which is the formula for *Linear Interpolation.*

It should be noted that only first order differences are involved in this formula. In Volume II more general formulae involving higher order differences will be obtained.

By inspection it is seen that the product $p.\Delta f_0$ represents the proportional part we considered earlier in the numerical example. The error in using this method is represented by the length PQ, as shown on the diagram. Such an error should not be neglected and in the first chapter of Volume II we show that the error will be less than $\pm \frac{1}{2}$ in the last digit of the calculated function value, provided that the second order difference $\Delta^2 f < 4$.

This condition can always be checked by inspection of the difference table. For example in the problem considered in §8.1.1, using the tangent tables,

		f	Δf	$\Delta^2 f$
x_0	tan 50° 30′	1·213 10		
			432	
x_1	tan 50° 36′	1·217 42		2
			434	
x_2	tan 50° 42′	1·221 76		

Thus $\Delta^2 f$ is 2 and the condition is satisfied, so linear interpolation is sufficiently accurate.

8.3 USE OF LINEAR INTERPOLATION FORMULA

When the value of the independent variable at which the function value is required does not give a simple proportional part of the interval then the formula (2) is the more convenient to use. We demonstrate the method in the following example.

Using the following difference table evaluate $f(2{\cdot}127)$

	x	$f(x)$	Δf	$\Delta^2 f$
x_0	2·1	0·61		
			48	
x_1	2·2	1·09		1
			49	
x_2	2·3	1·58		2
			51	
x_3	2·4	2·09		

The method can be split into two distinct parts

(i) Evaluating the interpolating factor 'p' using

$$x_p = x_0 + p.h$$

$$2{\cdot}127 = 2{\cdot}1 + p.(0{\cdot}1)$$

$$\Rightarrow p = 0{\cdot}27$$

(ii) Now evaluate $f(2{\cdot}127)$ using

$$f_p = f_0 + p.\Delta f_0$$

Then

$$f_p = 0{\cdot}61 + (0{\cdot}27)\,(0{\cdot}48)$$

Using a hand-calculating machine to perform this calculation, we can incorporate a check into this working.

The decimal point settings will be

$$\text{S.R. 2.,} \quad \text{C.R. 2.,} \quad \Rightarrow \text{Acc 4.}$$

Then by setting f_0 into the accumulator, clearing the S.R. and setting Δf_0 and adding into the accumulator we obtain $f_0 + \Delta f_0 = f_1$ which checks the setting of both f_0 and Δf_0. Whenever linear interpolation is performed it is possible to incorporate this check. Subtracting Δf_0 and completing the multiplication $p.\Delta f_0$ we have f_p in the accumulator, which in this case is $f_p = 0{\cdot}7396$.

This introduces another important consideration, namely, to what degree of accuracy is this result permissible? At this stage of the subject, since the function values were given to 2D, can we expect to obtain an intermediate value to 4D? We shall study this problem more fully in the chapter on Interpolation in Volume II. Here we conclude intuitively, that it is unlikely that we could obtain an answer to a higher degree of accuracy than that of the original data. Thus we round off our answer to two decimal places, giving

$$f(2{\cdot}127) = 0{\cdot}74(2\text{D}).$$

8.4 EXAMPLES

In the following examples, check, by means of a difference table to see if Linear Interpolation is permissible. If so, calculate $f(x)$ for the given value of x.

1 Given

x	0·12	0·14	0·16
$f(x)$	1·3777	1·3787	1·3800

Evaluate $f(0{\cdot}127)$

2. If we are given the following information concerning a function

x	12·1	12·2	12·3	12·4
$f(x)$	40·41	42·84	45·29	47·76

(i) What does the completed table of differences suggest?

(ii) Calculate $f(12{\cdot}264)$

3. If $f(x) = e^x$ then

x	3·70	3·71	3·72
$f(x)$	40·447	40·854	41·264

Calculate $e^{3{\cdot}706}$

4. Given

x	0·70	0·75	0·80
$f(x)$	17·86	17·93	18·00

Evaluate $f(0{\cdot}723)$

5. If

x	1·0	1·1	1·2	1·3
$f(x)$	−0·31	−0·10	+0·13	0·38

Calculate $f(x)$ when $x = 1{\cdot}136$

6. Given

x	−0·11	−0·10	−0·09	−0·08
$f(x)$	1·1821	1·1800	1·1781	1·1764

Calculate $f(-\,0{\cdot}0972)$.

9
Numerical Integration

9.1 INTRODUCTION

In work of a scientific or technological nature it is often found that physical quantities to be evaluated are obtained, in the first place, in terms of integrals or derivatives. For example:

$\bar{x} = \int xydx \div \int ydx$ (Centroid of plane area under

$\bar{y} = \frac{1}{2}\int y^2dx \div \int ydx$ the curve $y = f(x)$)

$\int pdv =$ Work done by a gas during expansion

$ds/dt =$ Velocity, (where $s =$ distance at time t)

In some cases the required values of integrals or derivatives can be found by analytical means but many more have to be found by some other method because:

(a) It may not be possible to express the result of integration in terms of known elementary functions,

(b) numerical integration may be easier or quicker than calculating from the function obtained by formal integration,

or

(c) no analytical expression for the function to be integrated (or differentiated) may be available; the available information being restricted to tabulated values.

Now any integral, no matter what physical quantities are involved, can be represented as an area. For example, $\int P\,dQ$ has a meaning only if P is a function of Q, and it can be evaluated only if:

(a) P is a known function of Q, or

(b) values of P are known at a sufficient number of values of Q.

In Fig. 9.1, the curve is obtained by plotting P against Q and has the equation $P = f(Q)$.

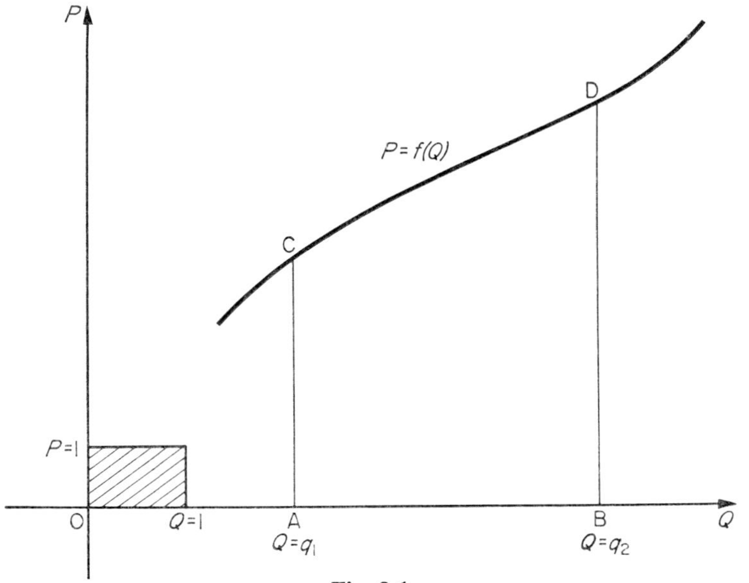

Fig. 9.1

The value of $\int_{q_1}^{q_2} P\,dQ$ is represented by the area* ACDB to a scale in which 1 unit is represented by the rectangle shown, whose length is 1 unit of the quantity Q and whose height is 1 unit of the quantity P.

Hence the value of $\int_{q_1}^{q_2} P\,dQ$ can be determined:

either

(a) by analytical methods in which case the value is

$F(q_2) - F(q_1)$ where $F(Q) \equiv \int f(Q)dQ$,

or

(b) by accurate drawing and measurement, (e.g. see planimeter §1.1.3), or

(c) by one of the numerical methods to be discussed in this Chapter.

Which method is used will depend upon the nature and any special features of the problem, and also to some extent upon the accuracy required in the result.

9.1.1 Exact data

If we know the exact values of the co-ordinates of a set of points there is an infinite number of different curves which all pass through the given points.

* Refer to any standard book on the Integral Calculus.

The only curve we wish to draw is the simplest or "smoothest" of all the possible curves, and intuitively we aim at drawing the curve whose curvature is as small as possible everywhere along its path.

We would therefore choose the full line in the diagram (Fig. 9.1.1) in preference to the dotted line because, of the two curves, it is the one which passes through all the points in the simplest manner.

Fig. 9.1.1

9.1.2 Approximate data

Tabulated values of a function are seldom known exactly. They are usually either rounded numbers, or have been obtained by measurements, and in either case will involve errors of unknown magnitude. Under these conditions the curve which is required will be the "straightest" possible curve which also passes as near as possible to each of the given points.

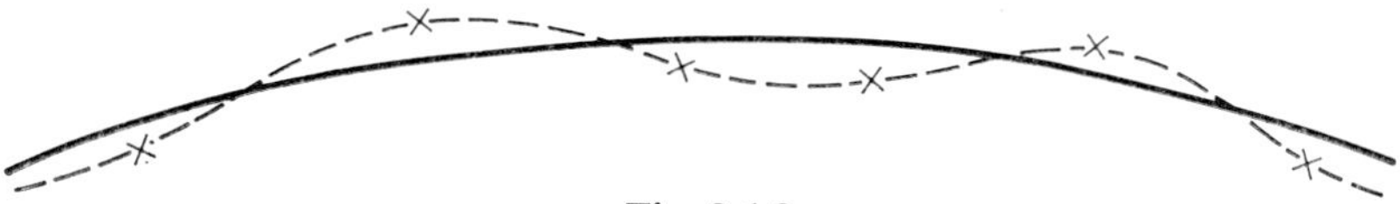

Fig. 9.1.2

Although the dotted line in the diagram (Fig. 9.1.2) actually passes through the given points, the full line represents the more probable graph of the function in most cases. Notice that the apparent errors in the positions of the given points are equally balanced on opposite sides of the chosen line. In general it is impossible to find the actual equation of a curve from numerical values of the function it represents. There are however methods of calculating the equation of a curve which agrees accurately enough with tabulated values. (See Vol. II chapter 7).

9.1.3 Polynomial approximations to functions

It follows from the above sections that interpolation and numerical integration and differentiation must all depend for their success on the choice of a suitable function which is to represent the data with sufficient accuracy. In elementary work it is found that polynomials are the most convenient to use for all these purposes, especially when each approximation need only remain accurate over a small range of values of the independent variable.

All the formulae for numerical integration used in this chapter are equivalent to the use of polynomial approximations to the function involved. One property of these polynomials is that they agree exactly with the tabulated values, that is, their graphs do pass through the points which represent the given data. We will therefore be using curves of the dotted type of Figure 9.1.2 rather than those of the solid line type, because they yield polynomials which can be used more readily. Curves of this solid line type will be referred to again in Volume II, Chapter 7.

Since many different polynomials can be chosen to represent any one function, there will be many different integration formulae available each of which will give results to some particular degree of accuracy.

The next section introduces the basic ideas of numerical integration, and uses an intuitive geometrical approach to the problem, A few simple, but widely used, elementary formulae will be derived in this way.

9.2 NUMERICAL INTEGRATION AND THE USE OF THREE SIMPLE RULES

The curve in Figure 9.2 is either the graph of a given known function of x, or is a curve obtained from numerical values of an unknown function of x as in Figure 9.1.1 or Figure 9.1.2. The problem to be discussed now is the problem of calculating the area NABM, i.e. the area bounded by the curve, the x axis, and the ordinates at $x = x_0$ (NA) and at $x = x_n$ (MB). (In the calculus notation this would be written as $\int_{x_0}^{x_n} f(x)dx$.

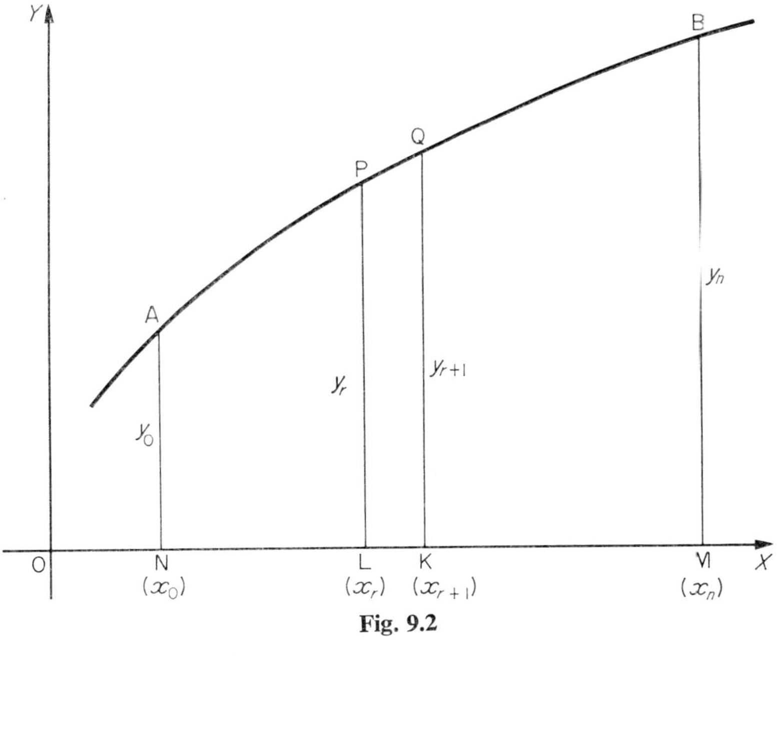

Fig. 9.2

Suppose the base NM of this area is divided into n equal parts, each of length $h(= \text{NM}/n)$ and that the values of x at the points of division are $x = x_0, x_1, x_2, \ldots, x_r, \ldots, x_n$. The ordinates $y = y_0, y_1, y_2, \ldots, y_r, \ldots, y_n$, at these points are either given in the table of values or can be read off from an accurately drawn graph, We now have the area divided into n strips all of the same width h. Methods of calculating the required area depend upon determining as accurately as possible the areas of these individual strips.

9.2.1 Trapezoid (or trapezium) rule

Consider any one of the strips such as LPQK where LP $= f(x_r) = y_r$ and KQ $= f(x_{r+1}) = y_{r+1}$. One first approximation to the area of this strip is obtained by assuming that the short arc PQ is very nearly straight, i.e. that the strip is a trapezium whose parallel sides are of lengths y_r and y_{r+1} at a distance h apart. Hence the area LPQK $\simeq (h/2)(y_r + y_{r+1})$. Applying the same method to every strip and adding the results gives:

$$\text{Area NABM} \simeq \frac{h}{2}(y_0 + y_1) + \frac{h}{2}(y_1 + y_2) + \ldots + \frac{h}{2}(y_{n-1} + y_n)$$

$$\simeq h(\tfrac{1}{2}y_0 + y_1 + y_2 + \ldots + y_{n-1} + \tfrac{1}{2}y_n)$$

Worked example 1(*a*)

$\int_0^{\pi/2} \sin x \, dx$ using the trapezoid rule.

The table gives values of sin x for $x = 0°\ (15°)\ 90°$

x	0°	15°	30°	45°	60°	75°	90° $= \pi/2$ radians
sin x	0	0·2588	0·5000	0·7071	0·8660	0·9659	1·0000

$$\tfrac{1}{2}(y_0 + y_6) = 0{\cdot}5000$$

$$y_1 + y_2 + y_3 + y_4 + y_5 = 3{\cdot}2978$$

$$3{\cdot}7978$$

$$h = \frac{\pi}{12} = 0{\cdot}261\,80$$

$$\therefore \int_0^{\pi/2} \sin x \, dx \simeq h(\tfrac{1}{2}y_0 + y_1 + y_2 + \ldots y_5 + \tfrac{1}{2}y_6)$$

$$= 0{\cdot}261\,80 \times 3{\cdot}7978$$

$$= 0{\cdot}9943 \ldots$$

Note. The correct result is 1, therefore the error involved in this example is $\simeq 0{\cdot}006$ or about 0·6 of 1%.

9.2.2 Mid-ordinate rule

Another way of finding an approximation to the area of the strip LPQK depends upon calculating or reading from the graph the length of the ordinate midway between LP and KQ (i.e. the ordinate at the mid point of LK). The length of this ordinate can be denoted by $y_{r+\frac{1}{2}}$ and we assume that this is nearly equal to the average height of the strip. Then the area LPQK is $\simeq$ equal to $h \,.\, y_{r+\frac{1}{2}}$ and by adding we have:

$$\text{Area NABN} \simeq h(y_{\frac{1}{2}} + y_{1\frac{1}{2}} + y_{2\frac{1}{2}} + \ldots + y_{n-\frac{1}{2}})$$
$$= h \times \text{(the sum of the mid-ordinates).}$$

Worked example 1(*b*)

$\int_0^{\pi/2} \sin x \, dx$ using the mid-ordinate rule.

Table of required mid-ordinates:

	0°	15°	30°	45°	60°	75°	90°($\pi/2$)
x	$7\frac{1}{2}°$	$22\frac{1}{2}°$	$37\frac{1}{2}°$	$52\frac{1}{2}°$	$67\frac{1}{2}°$	$82\frac{1}{2}°$	
$\sin x$	0·1305	0·3827	0·6088	0·7934	0·9239	0·9914	

sum of mid-ordinates $= 3{\cdot}8037$

$$h = \frac{\pi}{12} = 0{\cdot}261\ 80$$

$$\therefore \int_0^{\pi/2} \sin x \, dx = h \times \text{(sum of mid-ordinates)}$$
$$= 0{\cdot}261\ 80 \times 3{\cdot}8307$$
$$= 1{\cdot}0029 \ldots$$

Note. This shows an error $\simeq$ 0·3 of 1%.

9.2.3 Comparison of trapezoid and mid-ordinate rules

The two rules used above are equivalent to using a linear approximating polynomial since straight lines have been used to replace the sections of the curve. In one case each arc is replaced by its chord. In the other case a line through the top of the mid-ordinate and parallel to the chord is used. This is shown more clearly in Figure 9.2.3 which shows the arc PQR greatly enlarged in the case in which the arc is concave downwards. The same rules are now applied to the double strip, using the three consecutive ordinates LP $= y_r$, KQ $= y_{r+1}$, and

JR = y_{r+2}, and where the line SQT is drawn through Q parallel to the chord PR.

Using the trapezoid rule we have:

Area LPQRJ > Area of trapezium LPURJ

$$\text{i.e.} > 2h \times \tfrac{1}{2}(y_r + y_{r+2})$$

$$\text{i.e.} > h(y_r + y_{r+2}).$$

(The area PQRUP is ignored and an underestimate is obtained).

Using the mid-ordinate rule we have:

Area LPQRJ $< 2h \times y_{r+1}$

$$\text{i.e.} < 2h \times \tfrac{1}{2}(\text{LS} + \text{JT})$$

i.e. < Area of trapezium LSQTJ.

(The areas PSQ and QTR are ignored and an overestimate obtained).

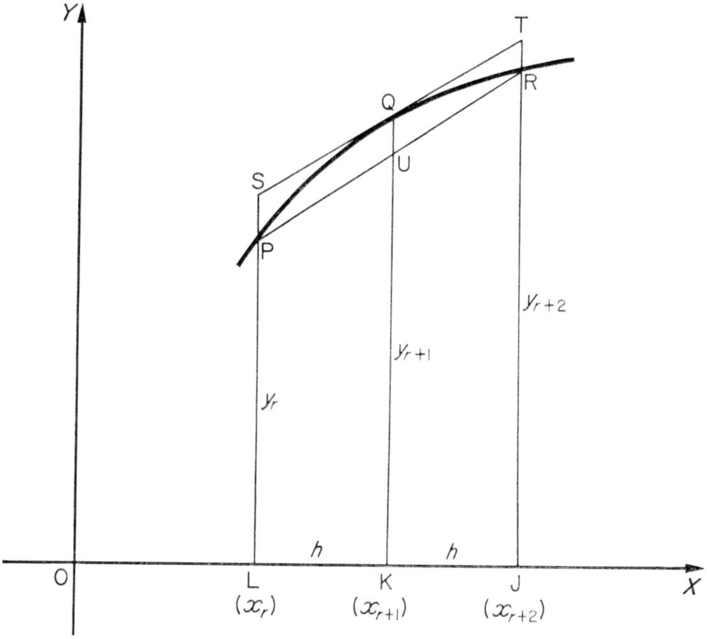

Fig. 9.2.3

When the arc is concave upwards the inequalities are reversed, but in either case the correct area lies somewhere between the values given by the two rules. This result is used in the next section to introduce a much more accurate method.

9.2.4 Simpson's rule

Let A_T = estimate obtained by using the trapezoid rule

$= \text{area LPURJ} = h(y_r + y_{r+2})$

and A_M = estimate obtained by using the mid-ordinate rule

$= \text{area LSQTJ} = 2h \, . \, y_{r+1}$

Then $A_M - A_T$ = area of parallelogram PSTR.

If we now assume that the area PQRUP is equal to two thirds* of the parallelogram PSTR we have:

The area PQRUP $= \frac{2}{3}(A_M - A_T)$

and the required area LPQRJ

$$\begin{aligned} &= \text{area LPURJ} + \text{area PQRUP} \\ &= A_T + \tfrac{2}{3}(A_M - A_T) \\ &= \tfrac{1}{3}(A_T + 2A_M) \quad \text{(Average of } A_M, A_T \text{ weighted 2:1)} \\ &= \tfrac{1}{3}h(y_r + 4y_{r+1} + y_{r+2}). \end{aligned}$$

This formula is exact whenever PQR is an arc of a parabola or a cubic† and when applied to other curves the accuracy obtainable is far greater than the simplicity of the formula would suggest. It is the equivalent of replacing each short section of the curve by a third degree polynomial approximation.

If the number (n) of strips used in subdividing the area NABM in Figure 9.2 is an even number we can apply this new rule to each pair of strips in turn and add the results.

Hence, area NABM

$$\begin{aligned} &\simeq \tfrac{1}{3}h(y_0 + 4y_1 + y_2) + \tfrac{1}{3}h(y_2 + 4y_3 + y_4) + \ldots \\ &\qquad \ldots + \tfrac{1}{3}h(y_{n-2} + 4y_{n-1} + y_n) \\ &\simeq \tfrac{1}{3}h(y_0 + 4y_1 + 2y_2 + 4y_3 + \ldots + 2y_{n-2} + 4y_{n-1} + y_n). \end{aligned}$$

Worked example 1(c)

Use Simpson's rule to evaluate $\int_0^{\pi/2} \sin x \, dx$.

Referring to the table of values given in Example 1(a) (see §9.2.1) we have:

$$\begin{aligned} &y_0 + y_6 && = 1{\cdot}0 \\ &2(y_2 + y_4) = 2 \times 1{\cdot}3660 && = 2{\cdot}7320 \\ &4(y_1 + y_3 + y_5) = 4 \times 1{\cdot}9318 && = 7{\cdot}7272 \\ & && \quad \overline{11{\cdot}4592} \end{aligned}$$

* This assumption is correct if PQR is an arc of a parabola, and in the same case ST is the tangent at Q.

† See section 9.3 below.

$$\therefore \text{Area} \simeq \tfrac{1}{3}h\,\{(y_0 + y_6) + 2(y_2 + y_4) + 4(y_1 + y_3 + y_5)\}$$
$$= \tfrac{1}{3} \times 0{\cdot}261\,80 \times 11{\cdot}4592 = 1{\cdot}000\,01$$
$$\text{i.e.} = 1{\cdot}0000 \text{ (to 4D)}$$

Correct value of the same area:

$$= \int_0^{\pi/2} \sin x \, dx = 1 \text{ (exactly)}$$

Note. The curve $y = \sin x$ is concave downwards all the way. Method (a) gives a result too small, error $= -\,0{\cdot}0057$. Method (b) gives a result too large, error $= +\,0{\cdot}0029$. Method (c) is in this special case correct to 4D, which is perhaps better than could have been expected.

Although more figures have been retained than appear justified they do show the expected ratios in the errors.

9.2.5 The use of different widths for different sections of an area

The number and width of the strips used in estimating an area, (i.e. the interval of tabulation required to evaluate the integral of a function) will depend upon the nature of the curve (or function). In many cases different widths may be used for different parts of the area. A simple example follows which illustrates this point.

Worked example 2

Evaluate $\int_1^{10} dx/x$

The table shows values of $1/x$ for $x = 1(1)10$ and the second and fourth differences. (See Chapter 5).

x	1	2	3	4	5	6	7	8	9	10
$1/x$	1·0000	0·5000	0·3333	0·2500	0·2000	0·1667	0·1429	0·1250	0·1111	0·1000
δ^2	—	3333	834	333	167	95	59	40	28	20
δ^4	—	—	1998	335	94	36	17	7	4	3

For a third degree polynomial to fit closely it is necessary that fourth differences must be small, and the table clearly shows that in this example a smaller interval must be used for the smaller values of x, while a larger interval could give reasonable accuracy for the larger values of x.

We could divide the range ($1 < x < 10$) of integration into the three sections $1 < x < 2$, $2 < x < 6$, and $6 < x < 10$ and using $h = \frac{1}{2}$, $h = 1$, and $h = 2$ respectively the calculation could be set out as below.

A simple known function is used in this example so that results obtained can readily be checked. We have $\int_1^2 (dx/x) = \log 2$, $\int_2^6 (dx/x) = \log 3$, $\int_6^{10} (dx/x) = \log (10/6)$ and $\int_1^{10} (dx/x) = \log 10$. Values of these have been included to enable comparisons to be made, thus showing that the same order of accuracy has been obtained in all three sections.

x	$1/x$	Multi-plier	$\frac{1}{3}h$	Correct value	Error
1	1·0000	1	1·0000		
1·5	0·6667	4	2·6668		
2	0·5000	1	0·5000		
			$4{\cdot}1668 \times \frac{1}{6} = 0{\cdot}6945$	$\log 2 = 0{\cdot}6931$	$0{\cdot}0014 \simeq 0{\cdot}2\%$
2	0·5000	1	0·5000		
3	0·3333	4	1·3332		
4	0·2500	2	0·5000		
5	0·2000	4	0·8000		
6	0·1667	1	0·1667		
			$3{\cdot}2999 \times \frac{1}{3} = 1{\cdot}1000$	$\log 3 = 1{\cdot}0986$	$0{\cdot}0014 \simeq 0{\cdot}13\%$
6	0·1667	1	0·1667		
8	0·1250	4	0·5000		
10	0·1000	1	0·1000		
			$0{\cdot}7667 \times \frac{2}{3} = 0{\cdot}5111$	$\log 10/6 = 0{\cdot}5108$	$0{\cdot}0003 \simeq 0{\cdot}1\%$
			$\therefore \int_1^{10} \frac{dx}{x} \simeq 2{\cdot}3056$	$\log 10 = 2{\cdot}3026$	$0{\cdot}003 \simeq 0{\cdot}13\%$

Worked example 3

The volume of a frustum of a right circular cone of height h whose ends have radii r and R is equal to $\int_0^h \pi y^2 \, dx$ where y is the radius at distance x from one end and where y is a linear function of x.

The volume can therefore be obtained exactly by Simpson's rule:

$$\text{Volume of frustum} = \tfrac{1}{3} \cdot \frac{h}{2} \cdot \left\{ \pi r^2 + 4\pi \left(\frac{R + r}{2}\right)^2 + \pi R^2 \right\}$$

$$= \tfrac{1}{3}\pi h \,(r^2 + rR + R^2)$$

Worked example 4

Find the area enclosed between the curves (1) $y = (3 - x)(x - 1)$ and (2) $y = x(x - 1)(x - 3)$ for $1 < x < 3$.

This area $= \int_1^3 (y_1 - y_2)dx$ where $y_1 - y_2$ is a third degree polynomial in x.

Hence the area can be obtained exactly by Simpson's rule as follows:

x	1	2	3
$y_1 = (3 - x)(x - 1)$	0	1	0
$y_2 = x(x - 1)(x - 3)$	0	-2	0
$y_1 - y_2$	0	3	0

$\therefore$ The required area $= \frac{1}{3} \cdot 1 \cdot (0 + 4 \times 3 + 0) = 4$ units.

9.2.6 Examples

1. Show that $y = (x + 1)(x - 2)(x + 3)$ and $y = (x + 1)(x + 2)(x + 3)$ intersect in two points only and use Simpson's rule to calculate the area enclosed between the two curves.

2. Tabulate values of $(\sin x)/x$ for $x = 0(\pi/6)\pi$ and hence estimate the value of $\int_0^\pi (\sin x/x)dx$.

3. Tabulate values of $\cos x \cosh x$ for $x = 0\,(0{\cdot}1)\,1$ using 6 figure tables, hence find an approximate value of $\int_0^1 \cos x \cosh x \, dx$.

9.3 ALGEBRAIC PROOF OF SIMPSON'S RULE

Referring to the Fig. 9.2.3 the points P, Q, R determine a parabola which passes through them, and in many practical examples the arc PQR of this parabola is a close approximation to the actual arc PQR of the given curve. There are also an infinite number of cubics which pass through the three points and one of these cubics can be chosen to provide an arc which gives a much better fit than the parabola. There is therefore very little error involved in assuming that the arc PQR is part of a cubic. Simpson's rule is based on this assumption.

The calculation is made much simpler by a transfer of the y axis, as shown in figure 9.3, where K becomes the new origin and the new co-ordinates of P, Q,

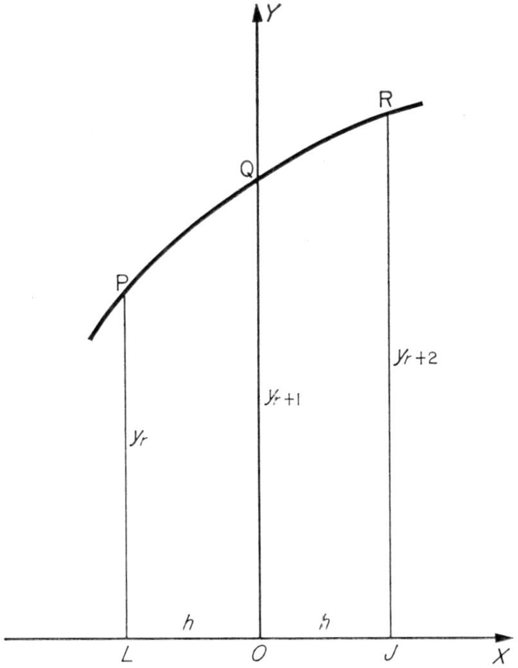

Fig. 9.3

and R become $(-h, y_r)$, $(0, y_{r+1})$ and $(+h, y_{r+2})$ respectively. We assume that the arc PQR is an arc of a cubic i.e. it has an equation of the form:

$$y = a_0 + a_1x + a_2x^2 + a_3x^3.$$

Because P, Q, R are points on this curve substituting their co-ordinates in the equation gives:

P is $(-h, y_r)$ $\quad \therefore y_r = a_0 - a_1h + a_2h^2 - a_3h^3$ (1)

Q is $(0, y_{r+1})$ $\quad \therefore y_{r+1} = a_0$ (2)

R is $(+h, y_{r+2})$ $\quad \therefore y_{r+2} = a_0 + a_1h + a_2h^2 + a_3h^3$ (3)

Adding (1) + (3) gives $y_r + y_{r+2} = 2(a_0 + a_2h^2)$ (4)

The area LPQRJ is given by:

$$\begin{aligned}\int_{-h}^{+h} y\,dx &= \int_{-h}^{+h} (a_0 + a_1x + a_2x^2 + a_3x^3)dx \\ &= [a_0x + \tfrac{1}{2}a_1x^2 + \tfrac{1}{3}a_2x^3 + \tfrac{1}{4}a_3x^4]_{-h}^{+h} \\ &= (a_0h + \tfrac{1}{2}a_1h^2 + \tfrac{1}{3}a_2h^3 + \tfrac{1}{4}a_3h^4) - (-a_0h + \tfrac{1}{2}a_1h^2 \\ &\qquad - \tfrac{1}{3}a_2h^3 + \tfrac{1}{4}a_3h^4) \\ &= 2a_0h + \tfrac{2}{3}a_2h^3 \\ &= \tfrac{1}{3}h\,(6a_0 + 2a_2h^2) \\ &= \tfrac{1}{3}h\,(4y_{r+1} + y_r + y_{r+2}) \text{ from (2) and (4) above.}\end{aligned}$$

which establishes the area of the *double* strip as approximately equal to:

The sum of the two outer ordinates and four times the central ordinate multiplied by one third of the *single* strip width

$$\text{Area LPQRJ} = \tfrac{1}{3}h\,(y_r + 4y_{r+1} + y_{r+2}).$$

The same argument and calculation can now be applied to each double strip in turn. Adding we have as in § 9.2.4 the formula:

Area divided into an *even* number of strips

$$= \tfrac{1}{3}h\,\{(y_0 + y_n) + 2(y_2 + y_4 + \ldots y_{n-2}) + 4(y_1 + y_3 + \ldots + y_{n-1})\}$$

or $= \tfrac{1}{3}h\,(y_0 + 4y_1 + 2y_2 + 4y_3 \ldots + 4y_{n-1} + y_n)$

9.3.1. Other rules for approximate evaluation of areas can be established by similar geometrical or algebraic methods and include:

(a) *The three-eighths rule*—for cases where the number of strips is a multiple of three:

$$\int_0^{3h} y\,dx \simeq \frac{3h}{8}\,(y_0 + 3y_1 + 3y_2 + y_3)$$

(b) *Weddles rule*—for an area divided into six strips:

$$\int_0^{6h} y\,dx \simeq \frac{3h}{10}\,(y_0 + 5y_1 + y_2 + 6y_3 + y_4 + 5y_5 + y_6)$$

Further information and other formulae of this type may be found in:

(a) Interpolation and Allied Tables—Page 70

(b) Shorter Six-figure Tables—Pages 352–3.

9.4 FURTHER METHODS FOR APPROXIMATE INTEGRATION

The formulae we have derived geometrically were originally devised with the specific aim of obtaining reasonable accuracy while at the same time only using the simplest possible multiples of the given ordinates, thus reducing the labour of calculation to a minimum. With the advent of modern calculating machines the necessity for minimising the labour involved is not nowadays so critical, and more sophisticated formulae can be conveniently used. Further methods of approximate integration using finite difference formulae are dealt with in Volume II. Chap. 4.

9.5 EXAMPLES

In the following examples, work to the number of significant figures provided by the tables you have available.

1. Find the area, in the first quadrant, bounded by the curve $y = (x + 5)(x - 1)(9 - x)$ and the x axis.

2. Tabulate values of $1 + \log_e x$ for $x = 3{\cdot}0\ (0{\cdot}1)\ 3{\cdot}6$ and hence evaluate $\int_{3\cdot0}^{3\cdot6} (1 + \log_e x)\, dx$.

3. Tabulate values of $\sqrt{(1 - \frac{1}{4}\sin^2 x)}$ for $x = 0\ (\pi/12)\ \pi/2$ and hence evaluate $\int_0^{\pi/2} \sqrt{(1 - \frac{1}{4}\sin^2 x)}\, dx$

4. Evaluate $\int_0^{0\cdot6} \sqrt{(x^2 + x + 7)}\, dx$.

5. If $y = \cos x$, tabulate values of y, xy and y^2 for $0\ (\pi/12)\ \pi/2$. Hence find the co-ordinates of the centroid of the area in the positive quadrant bounded by the curve $y = \cos x$ and the axes of co-ordinates.

6. By tabulating values of $\sqrt{\{1 - (x^2/25)\}}$ for $x = 0\ (0{\cdot}5)\ 5$ verify that the area of the ellipse $(x^2/25) + (y^2/4) = 1$ is 10π.

Answers to Examples

Chapter 1

Examples 1.10 Page 18

1 −31·252 **2** −1·390 **3** −1·326

4 −0·0001 **5** Final check sum −5·4787

Examples 1.13 Page 22

1 121·08 **2** 112·0054 **3** (a) 198·143 046; (b) 362·8800

4 − 9·2933 **5** 270·456 **6** − 43·239

7 467·655 28 **8** 3·215 36 **9** $6·0 \times 10^{-3}$; 13·3 (1D)

Examples 1.15.4 Page 27

1 135 **2** 10 067, rem. 32 **3** 11 433, rem. 10

4 4914 **5** 981

Examples 1.20 Page 37

1 6·7713 **2** 8·769 **3** 8·066

4 (a) 1×10^{-3}; 8·05 (2D); (b) 1×10^{-2}; 14·6 (1D), (c) $1·4 \times 10^{-3}$; 4·26 (2D)

5 $3·3 \times 10^{-2}$ **6** 1·370 37; 2×10^{-5} **8** 8·76

9 2·6457 ± 0·0003 **10** ± 0·004; 0·81 cu in (2D)

11 (a) 4, 0. (b) 3·59 **12** 2·28; 0·33; 0·145

Chapter 2

Examples 2.2.3 Page 45

1 (i) 2, (ii) 35, (iii) 126, (iv) − 7, (v) 29, (vi) 12

2 (i) 1, (ii) − 1, (iii) − 1, (iv) 1, (v) $f''(x) = -f(x)$

Examples 2.3.5 Page 52

1 (i) 8·565, (ii) 6·930, (iii) 2·618, (iv) − 0·028, (v) − 0·183, (v) 0·001, (vii) − 4·230

2 2·2 or 2·18 ± 1 in last place **3** 1·5 or 1·48 ± 1 in last place

4 − 65

Examples 2.4.5 Page 57

1 (i) 0·40, (ii) 0·99, (iii) 0·178 or 0·1776 $\pm$ 1 in last place, (iv) 1·4

2 (i) 5D, 0·401 53, (ii) 6D, 0·498 748 (use $\sin 2x$), (iii) 6D, 3·544 683 (iv) 6D, 1·264 145

3

x	1·0	1·05	1·10	1·15
$(1 + x^2)\sin x$	1·682 94	1·823 76	1·969 57	2·119 89
	1·20	1·25	1·30	1·35
	2·274 18	2·431 77	2·591 97	2·753 98
	1·40	1·45	1·50	
	2·916 93	3·079 90	3·241 86	

4

x	0	0·01	0·02	0·03
$(\cos^2 x - x^2)/\cos^2 x$	1	0·999 900	0·999 600	0·999 099
		0·04	0·05	
		0·998 397	0·997 494	

5

x	0·2	0·4	0·6	0·8
$2x + \log x$'	$-$ 2·818 88	$-$ 1·032 58	$+$ 0·178 35	1·153 71
	1·0	1·2		
	2·000 00	3·129 29		

Examples 2.5.1 Page 60

1 0·405 **2** 0·149 438 **3** 0·900 447

4 0·160 684 **5** 1·022 131

Chapter 3

Examples 3.7 Page 93

5 (a) $-\sqrt{3} < x < -1$ and $1 < x < \sqrt{3}$

5 (b) $0{\cdot}755 < x < \infty$.

5 (c) $0 < x < 1{\cdot}87$ and small ranges near $(5\pi/2)$, $(9\pi/2)$ . . . $-7{\cdot}8 < x < -1{\cdot}87$ and all negative values of x except those near $-(5\pi/2)$, $(9\pi/2)$. . . .

6 (a) Three, $x = 1$ $x \simeq 1{\cdot}4$ and $x \simeq 0{\cdot}8$

6 (b) Infinite, $x = 0, \pm 1{\cdot}2$ . . , $\pm(3\pi/2)$ (near), $\pm(5\pi/2)$ (near). . . .

6 (c) one, $x \simeq 9$ **6** (d) none.

6 (e) Infinite $x \simeq 1{\cdot}3$, $\pm 3{\cdot}5$, $\pm 2\pi$, $\pm 3\pi$, . . .

6 (f) Two $x \simeq 5$, $y \simeq -9$ and $x \simeq 5$, $y \simeq -9$.

7 (a) $x = -1{\cdot}447$, 1, 0·7413.

7 (b) $x = 0$, $\pm 1{\cdot}166$, $\pm 4{\cdot}604$, etc.

7 (c) $x = 9{\cdot}044$ **7** (e) $x = \pm 1{\cdot}307$, etc.

7 (f) $x = 5{\cdot}06$ $y = 9{\cdot}01$.

Chapter 4

Examples 4.3.3 Page 102

1 (a) 1·316 704 0, (b) 2·924 017 9, (c) 1·648 721 3

Examples 4.4.1 Page 102

1 3·0809 **2** (b) divergent, (c) divergent, (d) convergent

Example 4.4.2 Page 104

1 (a) $+ 0\cdot1$, (b) $+ 9$, (c) $- 3$, (d) $- 0\cdot3$

Examples 4.4.4 Page 107

1 1·068 **2** 1·404 **3** 0·5314

4 1·466

Example 4.5.1 Page 108

2 (a) $\left.\begin{aligned} x &= -1\cdot9266 \\ y &= -1\cdot8533 \end{aligned}\right\}$ (b) $\left.\begin{aligned} x &= 4\cdot7232 \\ y &= 5\cdot3685 \end{aligned}\right\}$

3 $x = 0\cdot5618$, $y = 1\cdot2173$

4 $x = 9\cdot1277$, $y = 5\cdot6594$, $z = 3\cdot6534$

Examples 4.8 Page 116

2 $\sqrt{(2)} = 1\cdot414\,213\,562\,4$
$\sqrt{(20)} = 4\cdot472\,135\,955\,0$
$\sqrt{(0\cdot3)} = 0\cdot547\,722\,557\,5$
$\sqrt{(2/3)} = 0\cdot816\,496\,580\,9$
$\sqrt{(\pi)} = 1\cdot772\,453\,850\,9$

3 $x = -0\cdot297\,370$

4 (a) $x = 0\cdot725\,84$, (b) $x = 1\cdot049\,91$, (c) $x = 1\cdot862\,81$

5 $1/\sqrt{(2)} = 0\cdot707\,106\,781\,186\,5$
$1/\pi = 0\cdot318\,309\,886\,183\,8$

6 $x = 11\cdot4172$ **7** $x = 0\cdot6367, -1\cdot4096$

8 $x = -2\cdot0653$

10 (a) m is between $\pm 2/9\pi$ (approximately), (b) $x = 2\cdot7798$

11 (a) Three roots if $+1 < m < +\infty$, one root if $-\infty < m < +1$
(b) $x = 0, \pm 4\cdot4999$

12 $\left.\begin{aligned} x &= 1\cdot2415 \\ y &= -0\cdot9704 \end{aligned}\right\}$ $\left.\begin{aligned} x &= 0\cdot6065 \\ y &= 0\cdot9193 \end{aligned}\right\}$

13 $x = 1\cdot147\,767$ **14** $x = 0\cdot7321$, $y = 1\cdot3791$

15 $x = 1\cdot523\,61$ **16** (a) $x = 2\cdot177\,42$ **17** $x = 2\cdot180\,32$

18 $x = 2\cdot0218$, $y = 2\cdot2164$

19 $x = -0\cdot1224, +2\cdot3552$

20 (a) 1·0202, (b) 1·793 173

Chapter 5

Example 5.1.1 Page 120

Self checking

Example 5.1.7 Page 124

Self checking

Example 5.2.1 Page 128

1

x	0·05	0·06	0·07	0·08	0·09	0·10
$f(x)$	1·625	1·656	1·689	1·724	1·761	1·800

2

x	− 1·0	− 0·8	− 0·6	− 0·4	− 0·2	
$f(x)$	2·000	2·048	2·224	2·576	3·152	
x	1·0	1·2	1·4	1·6	1·8	2·0
$f(x)$	14·000	17·488	21·584	26·336	31·792	38·000

3

x	0·6	0·7	0·8	0·9	1·0
$f(x)$	7·776	16·807	32·768	59·049	100

4

x	0	0·1	0·2	0·3	0·4	
$f(x)$	− 0·7	− 0·4036	− 0·0608	+ 0·3788	0·9656	
x	0·5	0·6	0·7	0·8	0·9	1·0
$f(x)$	1·7500	2·7824	4·1132	5·7928	7·8716	10·4000

5

x	0	0·1	0·2	0·3	0·4	0·5
$f(x)$	3	2·217	1·296	0·279	−0·792	−1·875
x		0·6	0·7	0·8	0·9	1·0
$f(x)$		− 2·928	− 3·909	− 4·776	− 5·487	− 6·000

Example 5.3.2 Page 132

1 $f(0{\cdot}8) = 12{\cdot}240$ **2** $f(0{\cdot}5) = 1{\cdot}250$ **3** $f(1{\cdot}2) = 15{\cdot}448$

4 $f(1{\cdot}6) = 5{\cdot}8528$ **5** $f(-0{\cdot}4) = -1{\cdot}5924, f(0{\cdot}4) = -3{\cdot}9764$

Examples 5.3.4 Page 135

1 $f(0{\cdot}14) = 10{\cdot}950\,548$ **2** $f(1{\cdot}5) = 74{\cdot}388$ **3** $f(2{\cdot}06) = 57{\cdot}504$

4 $f(0{\cdot}985) = 3{\cdot}4955$ **5** $f(1{\cdot}66) = 0{\cdot}7910$

Examples 5.3.6 Page 138

1 $f(0{\cdot}4) = 8{\cdot}744, f(0{\cdot}5) = 9{\cdot}500$ **2** $f(1{\cdot}2) = 17{\cdot}7712, f(1{\cdot}6) = 42{\cdot}1552$

3 $f(0{\cdot}03) = -3{\cdot}8119, f(0{\cdot}06) = -3{\cdot}6076$

Examples 5.5.5 Page 145

1 $x^3 + 2x^2 - 8x - 5$ **2** $4x^3 - 4x^2 + 2x + 7$ **3** $\frac{1}{10}(x^3 - 6x - 4)$

4 $2x^3 - 3x^2 + 3{\cdot}1x + 7$ **5** $x^4 - 2x^3 + 3x + 1$ **6** $2x^3 + x^2 + 7x + 6$

Chapter 6

Examples 6.6 Page 174

1 $x = 0{\cdot}665, y = 0{\cdot}429$ **2** $x = 1{\cdot}269, y = 1{\cdot}200$ **3** $x = 1{\cdot}268, y = -0{\cdot}042$

4 $x = 0{\cdot}8250, y = -0{\cdot}0317$

5 $x = 1{\cdot}144, y = 0{\cdot}328\ z = 0{\cdot}523$ **6** $x = 1{\cdot}133, y = -0{\cdot}867, z = -0{\cdot}600$

7 $x = 0{\cdot}73, y = -0{\cdot}94, z = 1{\cdot}05$ **8** $x = 5{\cdot}467, y = -0{\cdot}867, z = -0{\cdot}533$

9 $x = 1{\cdot}353, y = 2{\cdot}412, z = 3{\cdot}706$ **10** $x = 0{\cdot}82, y = 0{\cdot}56, z = 0{\cdot}71$

11 $x = 1{\cdot}68, y = 0{\cdot}50, z = 3{\cdot}40$

12 $x = -0{\cdot}106$, $y = -0{\cdot}842$, $z = -0{\cdot}969$

13 $x = 3{\cdot}233$, $y = -0{\cdot}123$, $z = 4{\cdot}548$ **14** $x = 3{\cdot}64$, $y = -1{\cdot}84$, $z = -9{\cdot}83$

15 $x = 0{\cdot}094$, $y = -0{\cdot}076$, $z = -0{\cdot}069$

16 $x_1 = 0{\cdot}8544$, $x_2 = -0{\cdot}6001$, $x_3 = 1{\cdot}0491$

17 $x_1 = 0{\cdot}1234$, $x_2 = 0{\cdot}2341$, $x_3 = 0{\cdot}3412$, $x_4 = 0{\cdot}4123$

18 $x_1 = 0{\cdot}27$, $x_2 = 0{\cdot}24$, $x_3 = -0{\cdot}55$

Chapter 7

Examples 7.1.2 Page 180

1 $x^2 + 4x + 5$ r. 15 **2** $x^3 - 3x^2 + 3x - 5$ r. 16

3 $3{\cdot}5x - 11{\cdot}6$ r. $47{\cdot}7$ **4** $x^2 + 2x + 8$ r. 45

5 $\frac{1}{3}(4x^2 - 8x + 22)$ r. -51 **8** 19·08, 13·75

9 1·5646 **10** $-0{\cdot}623$

Examples 7.1.4 Page 184

1 $2x^2 - 4x + 11$ r. $-24x - 14$ **2** $3x^2 + 5x + 8$ r. $x - 16$

3 $2x^2 + 6x - 2$ r. $17x + 1$ **4** $8x - 27$ r. $152x - 140$

5 $0{\cdot}5x^2 - 1{\cdot}75x + 3{\cdot}625$ r. $-18{\cdot}75x + 14{\cdot}75$

6 $x^3 - 2x^2 - 6x - 13$ r. $-30x + 33$

7 $2x^3 - 6{\cdot}5x^2 + 14{\cdot}75x - 34{\cdot}125$ r. $158{\cdot}375x - 142{\cdot}5$

Examples 7.1.6 Page 186

1 $-10 + i\sqrt{(3)}$ **2** $125 + i\,245\sqrt{(2)}$

6 $-23{\cdot}514 + i\,19{\cdot}385$ **7** $-0{\cdot}003 + i\,0{\cdot}001$

Examples 7.3.1 Page 190

1 0·099 019 5, $-10{\cdot}0990$ **2** $-4{\cdot}5 + i\,0{\cdot}866\,025$, $-4{\cdot}5 - i\,0{\cdot}866\,025$

3 0·000 124 999, $-16{\cdot}0001$ **4** 0·000 833 391, 11·9992 **5** $2 \pm i\,2{\cdot}449\,49$

6 3·8 repeated **7** 0·000 500 03, 9·9995

Examples 7.4.1 Page 195

1 $-1{\cdot}247$, 0·445, 1·802 **2** 2, $-0{\cdot}414\,214$, 2·414 214

3 $-0{\cdot}732$, 0·5, 2·732 **4** $-1{\cdot}2695$, $1{\cdot}6348 \pm i\,0{\cdot}6916$

5 1·383, $-2{\cdot}191 \pm i\,0{\cdot}509$ **6** $-4{\cdot}472\,14$, 2·236 07 repeated

Examples 7.5.1 Page 195

1 1·50, $-5{\cdot}56$, $1{\cdot}47 \pm i\,2{\cdot}49$

2 $-1{\cdot}00$, 2·50, $7{\cdot}30 \pm i\,2{\cdot}24$

3 0·186, $-2{\cdot}686$, $-0{\cdot}5 \pm i\,1{\cdot}118$

Chapter 8

Examples 8.4 Page 200

1 1·3781 (4D)

2 (i) The given function values are exact. (ii) 44·41 (2D)

3 40·691 (3D) **4** 17·89 (2D) **5** − 0·02 (2D)

6 1·1794 (4D)

Chapter 9

Examples 9.2.6 *Page 212*

1 $5\frac{1}{3}$ **2** 1·852 3 0·9667

Examples 9.5 *Page 214*

1 $853\frac{1}{3}$ **2** 1·3155 **3** 1·4675

4 1·6341 **5** (0·5708, 0·3927)

Bibliography

ALLEN, D. N. DE G. (1954) *Relaxation Methods in Engineering and Science.* McGraw-Hill, London.

British Standard Letter Symbols, Signs and Abbreviations, BS 1991, Part 1 (1954). British Standards Institution, London.

British Standard Flow Chart Symbols, BS 4058, Part 1 (1966). British Standards Institution, London.

BUTLER, R. and KERR, E. (1967) *An Introduction to Numerical Methods.* Pitman, London.

COMRIE, L. J. (1966) *Chambers Shorter Six Figure Tables.*

COURANT, R. (1937) *Differential and Integral Calculus*, Vol. 1, 2nd edn. Blackie, Glasgow

FOX, L. (1957) *The Numerical Solution of Two-Point Boundary Problems in Ordinary Differential Equations.* Pergamon, Oxford.

HARTREE D. R. (1958) *Numerical Analysis*, 2nd edn. Oxford University Press.

Interpolation and Allied Tables (1955) H.M.S.O.

REDISH, K. A. (1961) *Introduction to Computational Methods.* English University Press.

WHITTAKER, SIR E. and ROBINSON, G. (1944) *Calculus of Observations*, 4th edn. Blackie, Glasgow.

Index